The Heavenly Time Machine

Essays on Science and Torah

The Heavenly Time Machine

Essays on Science and Torah

by Morris Engelson

Publisher's Cataloging-in-Publication
(Provided by Quality Books, Inc.)

Engelson, Morris
The heavenly time machine : essays on science and Torah / Morris Engelson. – 1st ed.
p. cm.
Includes bibliographical references and index.
LCCN: 2001126098
ISBN: 0-9642870-0-5

1. Bible and science. 2. Judaism and science. 3. Science – Philosophy. I. Title.

BS650.E54 2001 296.3'7'5
QB101-200180

Published by Joint Management Strategy (JMSC)

P.O. Box 25170, Portland, OR 97298

www.pcez.com/~jmsc

For publisher: jmsc@pcez.com

For author: m.engelson@ieee.org

To my grandchildren.

You are the next link in the chain.

Contents

Acknowledgements

I owe much to many people for their help and encouragement in preparing this book. Some of this help was given to me recently as I was preparing the manuscript. But much goes back in time to the point that I no longer recognize from where or how certain ideas came to me. This troubles me as I would like to publicly acknowledge my debt to all. Thank you, even if I can no longer identify you. Here are three examples from the past that I can identify.

The book *Challenge*, edited by Aryeh Carmell and Cyril Domb, has been a classic in the field of discourse between science and Torah since its publication in 1976. I have read and studied this work numerous times over the last two decades, and I know that it influenced the writing of this book in many ways. I discuss, for instance, the idea of a tree which is simultaneously three days old and three hundred years old, in chapter one. I am certain that this is an outgrowth of ideas presented in *Challenge*. This volume belongs in the library of all who are seriously interested in this field of study.

I was present at a celebration of completion where my friend, Baruch Broderson, gave a talk on Tractate *Rosh Hashanah*, to mark the completion of this section of Talmud in the *Daf Yomi* study program. Among other things, he noted the phenomenal accuracy to which the length of the lunar month was known. The reader will find a discussion on this matter in chapter seven. May Baruch be granted the merit to finish this difficult course of study which takes up several hours per day, every single day, for a period of over seven years.

Near twenty years ago my young son, Shlomo, asked me the question that all fathers dread. No, that is not the one. That one would be easy. Rather, he wanted an explanation of how the age of the universe is given as less than 6000 years according to Torah, and billions of years according to science. I was thrown into a panic because I had no acceptable answer. But I knew that there had to be one, and I started to investigate the matter. My investigations led to a result of 14.23 billion years, as discussed in this book. Now the reader can understand why I am partial to this result in preference to other formulations.

We all know that a tiny pinch of spice can transform a good meal into one that is superb. So too with skilled editing. We are all the beneficiaries of the, light-touch, editing of university professor and editor, Gardner Mein. I am grateful for his subtle improvements that make it appear as if I knew how to write. Likewise, the vehicle that contains the writing has benefited from the clean-looking layout and pleasing arrangement of pages and font provided by Dale Aufrecht of Technidata Services.

The first person that I told about this project, after my immediate family, was my rabbi, Rabbi Leonard Oppenheimer, spiritual leader of Congregation Kesser Israel. It was a smart move on my part. I now had a skilled research assistant at my disposal. He unfailingly knew the answers to my questions immediately upon being asked. Not only that, but he also suggested areas for me to look into that I would otherwise not have considered. The reader will find a good example in the footnote to chapter one. May he continue with a productive future of leading his people in the path of Torah.

A serious concern that I had from the outset was how this work might be perceived by those who teach and expound on the Torah on an ongoing basis. It goes without saying that any errors this book contains, whether factual or in analysis, are solely the responsibility of this author. But my concern about how matters of Torah are presented in this book was much alleviated when Rabbi Mordecai Kornfeld agreed to review the manuscript. Founder and Rosh Kollel of *Kollel Iyun Hadaf* and of the *Dafyomi Advanced Forum*, author of *Torah from the Internet* and other publications, and author of numerous Torah lessons available on the internet, Rav Kornfeld has honored me by spending his valuable time on these essays. His far-ranging comments, not only on matters of Torah, but on all aspects of my writing have made an important contribution to this project. May the time he spent on this endeavor be counted to his merit as if spent in the pure study of Torah. The following web site and email addresses will get the reader in touch with opportunities to learn Torah under the direction of Rabbi Kornfeld: www.dafyomi.co.il, www.shemayisroel.co.il, daf@dafyomi.co.il, ask@dafyomi.co.il, info@dafyomi.co.il, and kornfeld@virtual.co.il.

The first people who found out about this project were my immediate family. Knowingly, and sometimes unknowingly, as explained previously in the example with my son, my children Shlomo, Rivkah and Aviva made important contributions to this work. My wife, Janett, especially, is a partner to this work in the full sense of the meaning of an *ezer k'negdo*, as discussed in Chapter 9. This work might not have been written without my family's help and encouragement, and it certainly would have suffered for lack of their contributions.

Finally, one cannot write words of Torah without acknowledging the Source of the Torah itself. I long debated with myself whether I should tell here the story of how this book came to be written. I do not want to give the impression of being a superstitious person who is looking for omens. And yet the story that follows seems to have so many important lessons that I have chosen to tell it. My compromise is to just tell the story and stay away from any editorializing as to its meaning.

I am a consultant in a highly specialized technical field, with a busy work schedule. I am fortunate to be fairly well-known in the small community that uses my services, so I need not spend much time to find new clients. This is good because I dislike the effort, and am inept at finding new clients. One day something happened and I suddenly started losing my clients.

The first surprise came from a U.S. government agency that canceled a consulting trip just two weeks before it was scheduled to occur. They had no budget due to a dispute between the President and the Congress respecting a spending bill. Not only did I not receive the income that I was expecting, but I was also stuck with a costly non-refundable airline ticket. Shortly thereafter, another client canceled because a winter storm had damaged their facility. These are perfectly understandable happenings. Budget arguments in the U.S. government are not unknown. Storms are not unknown. Other cancellations followed. And these too seemed perfectly reasonable on an individual basis. But the result was most serious when taken in the aggregate. I had no work and I had no income. But I did have time. One of the things that I did with my time was to indulge in my hobby of researching matters of science and Torah, and I started to write down some of my ideas. The other thing I did with my time was to try and find some new clients.

Finding new clients is not easy, especially if you are not good at it. My first attempt was to send out a costly mailing to an area of the country where my specialty should have been in high demand. This yielded a few small clients that barely covered my canvassing costs. I asked these people why they would not use me on a more extensive basis. Business was great. Surely they could afford it? Well, that was the problem. Business was so great that they had no time to spend with me, even though they knew that I could provide an important contribution for the future. I now knew what to do. Go after organizations where business was bad. These people had plenty of time to spend with me. But, no. I was wrong again. Who spends money on consultants when you are worried about layoffs?

And so it went. My existing clients were gone, and I could not find any new ones. The only bright spot was that I seemed to be making real progress in my playing around with the science and Torah speculations. And then I got a phone call. The caller told me that he had a serious technical problem and needed my help desperately. When could I get on a plane and start to work on the matter? But I realized I did not want to work on this, or any other, consulting problem. I wanted to spend my time developing my ideas about the relationship between science and Torah. I decided to try and solve the problem immediately on the phone. I asked a series of questions and simultaneously made a number of back-of-envelope computations. I outlined a series of proposed experiments and suggested remedial actions depending on what they found. A week later this non-client called back with the results of the experiments. I outlined what he should do, and the case was closed. The client was delighted. His problem was solved and his organization had no expenses. I was also delighted. The problem was solved, and I did not have to take time away from my efforts at better understanding the relationship between science and Torah. And then I knew. I was not playing around. I was not indulging in speculations. I was writing a book.

But wait a minute. What about clients? What about income? The way I've come to look at it is, this is not my responsibility. I tried to find new clients, but I did not succeed. As for income, that comes from a higher Source than myself. My job is to use my time and my talent and the little bit of insight that I seem to have developed to pursue my explorations of science and Torah. The rest will eventually come to me, if I am worthy. I am no longer agitated about clients and income, and am able to spend uninterrupted months working on my book. A first draft is finished, and guess what? Suddenly the phone is ringing, emails start arriving and I am working again. It is as if the missing months never happened. Except that I have a book.

Now the book is finished, and I read the final draft. It seems a credible product. Did I really write all that? Where did I get these ideas? I do not know. But that is the scientist speaking. In fact I do know. I am reminded every day when I say these words:

You graciously endow man with wisdom

and teach insight to a frail mortal.

Blessed are You, Hashem,

gracious Giver of wisdom.

Morris Engelson
Sivan, 5761
June, 2001

Foreword

Dear reader. I am pleased that you are reading this book, and hope that it includes some material worthy of your attention. Nobody writes a book with the objective of being misunderstood. Certainly it is important to the author that the content be understood as to what the book is about. But in this instance, I feel it even more important to explain what this book is not about.

First, it is important to understand that this book does not teach Torah. It includes statements about Torah, but only in an incidental way. So the reader will get from this book a very incomplete picture of what Torah is about, at best. Anyone looking to learn Torah needs to look elsewhere. There are a number of reasons why it is not my intent to teach any Torah here, not the least being that I am not qualified to do so. Anyone who cannot study a page of Talmud without the help of dictionaries and translations, as is the case with me, has no business pretending to be a scholar in this field. Torah, in its entirety, is a most demanding field of study. The word Torah, in the subtitle of the book, is not intended as an alternate way of saying religion, or scripture or Bible. Torah is Torah, and there is no other single word equivalent. I mean Torah in all its aspects as understood in Orthodox Judaism. This includes both the oral and written Torah in the original language, with the precise variations in spellings. Not only spelling, but spacing of words and letters is important. The tiniest difference can have the largest impact. We must also include the wisdom hidden within the pronunciation accent marks and decorative crown embellishments, known as *tagin*. All of this, and more, is part of Torah.

Rabbi Munk notes on page 47 of *The Wisdom in the Hebrew Alphabet*, that "Concealed in the letters of the Torah, in their combinations, crowns, and shapes are the laws of physics, ... music, ... mining ..., ethics." This is why the danger of presenting a distorted picture of what Torah is, and is not, is very real. I again caution the reader to not mistake what I have done in this book as teaching Torah. Rashi, Maimonides, Nachmanides, Rabbi Dessler, to name some of the people quoted in this book, teach Torah. I am not qualified to teach Torah, and am not trying to do so.

Another matter that calls for clarification is the choice of elements of Torah, and of science as well, that I have chosen to present. Torah is vast and science is vast. Why have I chosen to dwell at length on the relatively obscure scientific idea known as the anthropic principle, to the exclusion of, say, conservation of energy? The reason is that I am acting as an advocate for a certain idea, and I have chosen those aspects of science and Torah that help me make my point. And yes, I have chosen to ignore those ideas in science and Torah that work against my point. This book is not a record of a search for truth. It is an explanation of my belief. I had my mind made up before I started. I say that the Torah is true, and science in its way is also true. Both teach truth, and both can coexist together and help and complement each other. That is what I am trying to show in these essays.

There are a number of Torah-based suggestions for the age of the universe. Most are better known than the formerly secret computations by Kaballah master, Rabbi Isaac of Akko. But I choose here to quote Rabbi Isaac for the simple reason that his numbers are in accord with current science-based results. This is nothing less than the fallacy of circling the target after the arrow hits it, as discussed in the essay on probability mathematics. Such behavior can be rightly criticized as not proving the truth of science or of Torah. Furthermore, what gives me the right to use this particular answer in preference to the several alternate answers that Torah gives us? Indeed, does not the fact that there are multiple, sometimes conflicting, answers invalidate all of Torah?

First permit me to state in the strongest terms that I am not seeking to prove the truth of Torah by scientific means. I know of no way to do this, and am convinced that it cannot be done. There are two primary reasons for this conclusion. One reason is that Torah assures us, as discussed in these essays, that ultimately the workings of the universe will be understood as part of a "natural" whole. We will one day have a unified field theory that describes the whole universe in a few succinct equations. Needless to say, this author does not accept the position that this invalidates the Torah. But a scientist could argue that there is no need for Torah once these equations become known. The other reason is given by science popularizer and fiction author, Isaac Asimov, in an essay in the April, 1975 issue of *Science and Science Fiction* magazine. Asimov says that "when Believers base an argument for the existence of G-d on scientific findings, they are calling upon the enemy, so to speak."

It is the purpose of my essays to argue that science and Torah need not be enemies, though many scientists, including Asimov, take the view that science and religion cannot coexist. To such scientists, any assertion, that science can prove the truth of Torah must also include the proposition that science has the right to prove the falsity of Torah. And who will be the judge of the result? The scientists, of course. But no one who takes Torah seriously would be willing to accept any judgment on the truth of Torah on the basis of conclusions from a Torah-ignorant and hostile scientist.

Look at the words used by Asimov, whose writings make it clear that he does not accept the validity of the Bible, "Believers base an argument…." "Belief" according to my dictionary involves "a state … of trust or confidence." Not proof, but "trust." Asimov will not accept my word on trust, no matter what science-based argument I make. He wants proof that there is no scientific alternative. But the Torah assures us that there will always be an alternate science-based explanation. This is how it has to be so that we may have free will to believe or to not believe. No, my essays are only intended to argue that science and Torah need not be enemies, and that science cannot falsify Torah. One may choose to believe or not believe in Torah for any number of reasons, but scientific proof is not involved in the choice. At most I am seeking to show that modern science does not hinder belief in Torah. In this respect, I should call the reader's attention to the elegant little book by Kelemen on the topic *Permission To Believe*, which deals with the relationship between Torah, science, history and a number of other fields of study.

Let us now consider the validity of the evidence that I have chosen to use. Science is a work in progress. Advances follow from the free competition among a number of differing theories. Ultimately most conjectures are shown to be incorrect, and only some few or, ideally, one is left intact. Until that happens, however, we are free to follow the inference of whichever theory we prefer. No one would consider this improper. But what about Torah, which is supposed to represent absolute truth? Surprisingly, there is much leeway in matters which have no impact on everyday practical results. Take, for example, the use of electricity on the Sabbath. Is the use of electricity among the thirty-nine categories of forbidden labors, or not? The answer is not obvious, and it takes considerable Torah knowledge to understand this matter. That is why it caused such misunderstanding for Professor Feynman, as we will see in the first chapter of this book. But the matter is quite clear after being analyzed. Electricity use falls within several forbidden work categories, including the kindling of fire, and it may not be used on the Sabbath. There is only one definitive answer in this practical matter. But, as explained by Rabbi Kaplan in his *Immortality, Resurrection, and the Age of the Universe*, "there is no one binding opinion" on a matter such as the age of the universe, which has no immediate impact on our lives. "Any opinion found in a recognized Torah source is acceptable."

The Torah talks to us on many levels, and even seemingly opposing opinions, derived from the Torah by competent authorities, are in the category of "both are the words of the living G-d." A person of my level of Torah learning, and even someone several rungs higher, should refrain from claiming novel interpretations not previously expounded. But I am free to choose any material from among the recognized authorities to help me illustrate a point.

Finally, I would like to address the question of whether science and Torah are, indeed, historical enemies, as some would have us believe. Let us look at a few historical examples.

Isaac was born over 3700 years ago, in the year 1711 BCE. This is 400 years prior to the exodus from Egypt and the giving of the Torah. At that time an Egyptian priest, named Ahmes, wrote a papyrus document which is now part of the Rhind collection in the British Museum. This document, titled "Directions for Knowing all Dark Things," deals with the mathematics of certain kinds of fractions and the computation of areas. Some of the problems discussed are not easy ones. For instance, what fractions with unity numerator when added together yield a sum of 2/29? Try for a solution before checking the answer at the end of this essay. The point I am trying to make is that this was highly advanced science in its day, and it was entirely in the hands of the priests. This was esoteric knowledge to ordinary people, involving "dark things." Mathematics, astronomy, and science in general, all were part of the religion, and closely guarded. Science and religion were the same, and there was no conflict. Then came the Torah, 400 years later, and it clearly pointed out how wrong it is to worship human beings, or the products of human effort, or natural objects such as the sun, as the ancient Egyptians did. We have a conflict involving Torah and sun worship. Science is involved to the degree that it was part of the Egyptian religion. But there is no conflict between science and Torah, *per se*.

Let us now take a quick trip in a time machine to a time only 400 years ago, as opposed to nearly 4000 years ago, when Ahmes wrote his treatise on fractions. All are familiar with the story of Galileo Galilei who was forced to recant the "heretical" notion that the sun and not the earth is the center of the solar system. Galileo knew that an accusation of heresy was a serious matter, and he wisely decided to live. Not so his predecessor, astronomer and philosopher Giordano Bruno, who refused to recant, was found guilty and was burned to death in 1600. Talk about conflict. How much worse can it get than for people to be killed? But what does the above have to do with Torah? Or with science, for that matter, given the many people who died for "crimes" not in any way related to science? The Inquisition was responsible for many horrors, including the death of Jews who were killed for their adherence to Torah. No, we do not have a historical example of conflict between science and Torah here.

Clearly science and theology had differences on the facts in Galileo's time. But science was not opposed to the faith aspects of the Bible even then, given that scientists accepted the idea of a "Creator." Rather it was science which was under attack by religion. A century later, we find that theology was incorporated into science at the highest levels. A good example is the Leibniz-Clarke debate. That Samuel Clarke, who was a theologian as well as philosopher and scientist, would use theological arguments is no surprise. But that Leibniz, whose version of the calculus we use today, would do so seems surprising. Yet that is exactly what Leibniz did, and his position was accepted as perfectly cogent. The debate, conducted by exchange of letters, revolved around the question of absolute versus relative space and time. Clarke championed the Newtonian position that all was absolute, while Leibniz argued that all was relative. The debate was cut short by the death of Leibniz in 1716. Indeed, the issue is not yet settled today, as scientists continue to ponder the meaning of space and time in light of relativity theory and quantum mechanics.

Leibniz opens the debate by an attack on the notion of absolute space because this idea implies that space is an organ which G-d makes use of to perceive things by. Obviously G-d has no need of such help, hence there is no absolute space. Clarke replies that G-d is certainly present in all of space. How is it possible for such space to not be absolute? Back and forth it goes, with arguments about the nature of space and time. Throughout, we have the unstated assumption, fully accepted by both individuals, that the existence of G-d is a given. Hence, any science that runs counter to this notion must be wrong. The debate was officially never resolved. But all accepted that Clarke was correct, given the great success of Newtonian physics. Paradoxically, the apparent success of Clarke, the theologian, led to a serious assault by Laplace on belief in G-d a century later.

Pierre Simon de Laplace (1749-1827) was a nimble fellow. He managed to prosper through the French revolution and its reign of terror, through the age of Napoleon and monarchical periods, while many of his contemporaries, literally, lost their heads. He was happy to pledge allegiance to whomever was in charge, and he likewise did not shrink from "borrowing" scientific results from his coworkers without attribution. Possibly he got away with this because he was undoubtedly a scientist of the first rank. He was considered only second to Newton in his time. It was Laplace who showed that the solar system will not lose its cohesion in the future, by taming the famous three-body problem – a problem, in gravitational mechanics, that is not fully solved to this day. He was a master in the analysis of Newtonian mechanics, and a staunch believer in its implications.

Laplace believed completely in the supremacy of science and a deterministic universe. He claimed that if we knew the exact position and velocity of every particle in the universe at any particular time, then we could work out all the past and future of the universe. When Napoleon asked why Laplace did not mention G-d in any of his writings, Laplace responded that he had no need of that hypothesis. This was a serious challenge, and its effect lasted for the next hundred years. But no more. Since the early twentieth century, we have been compelled to deal with relativity and quantum mechanics, and this changes everything. The hypothesis that Laplace discarded can no longer be ignored.

Human history is full of conflict between peoples and ideas. But none of it is related to any fundamental differences between science and Torah, and one sees this if one digs beneath the surface of initial appearance. It seems to me that there is no inherent historical experience that requires science and Torah to be enemies, the comments by Laplace, and similar comments by other scientists notwithstanding. Stephen Hawking, the Lucasian Professor of Mathematics at Cambridge University, a post once held by Newton, is not looking to exchange science for religion. But honesty compels him to say that given our current state of science he finds no rational way to explain the universe, except "as evidence of a Divine purpose in creation" (page 125, *A Brief History of Time*). With this, the Torah fully agrees.

Thank you for joining me in this journey to explore the relationship between science and Torah.

Answer to question about fractions:

$$2/29 = 1/24 + 1/58 + 1/174 + 1/232.$$

Chapter 1

Who's on First?

Modes of Interaction Between Science and Torah

I vaguely knew the name of Richard Feynman and his work in the same way that I am aware of other Nobel Laureates. This changed when I chanced to be present at the showing of a motion picture of Feynman's first lecture on *The Character of Physical Law*, presented at Cornell University in 1964. I was totally captivated and made sure to view the showings of all seven lectures. The power of these lectures, presented extemporaneously, was extraordinary. This was not just a superior scientist, but the greatest science teacher that I had ever been exposed to. I immediately purchased the transcription of these lectures. And I have gone through this little book many times, each time finding something new and marveling at the extraordinary genius of the presentation. I purchased the Feynman books on physics, and finally understood certain concepts that I had not really understood from my university course work. If it had Feynman's name on it, I wanted it. And I was never disappointed. I came to understand why some of his contemporaries claimed that Feynman was the smartest man who ever lived. And if not of all time, then certainly of the twentieth century. The most that the doubters would argue is that possibly Einstein should come before Feynman. No wonder his biography is simply titled *Genius*.

Not only was Feynman considered a genius in physics and in teaching, but also in understanding people. He loved to expose the stuffed-shirts and phonies among us. And he was very good at it. Numerous stories about his prowess to virtually read minds circulate among the cult-like Feynman-based literature and television programming.

With that background, it comes as no surprise that I was happy to receive as a present, in late 1986, the book *Surely You're Joking Mr. Feynman*. I now had something enjoyable to read on an upcoming air trip. The book is humorous, as is implied by the title. But it is also sad and serious too. Feynman talks about serious matters, but presented in the special humorous way that is the hallmark of his style. The book is a collection of loosely interconnected adventures and incidents in Feynman's life. Mostly we deal with an interaction between Feynman and other people. And mostly the other people do not end looking very good. Once Feynman gets you in his laser-like sight, it is impossible to escape till he strips you bare.

So here I am reading the book and silently chuckling to myself at the seeming supernatural insight of superhero Feynman. And then I come to a chapter titled *Is Electricity Fire*? Suddenly the book is no longer funny. I am so upset that I can't sit still. But there is no choice, because I am strapped in a seat on an airplane. "Sorry sir, you cannot go for a walk outside." I feel terrible. That which I value is being destroyed. Torah is insulted, Torah study is being ridiculed, and my hero Feynman is making a fool of himself. How is it possible for genius Feynman to be so confused?

Is electricity fire?

I found the fire story rather strange. Feynman is in New York in the mid nineteen-fifties for a conference, and he is mysteriously lodged at the Jewish Theological Seminary. He meets some of the students, whom he believes to be young Orthodox rabbis. They meet and they talk. Feynman believes that he is equipped to deal with the various topics because he is Jewish and he has the "background," as he puts it. The description of what happened is highly compressed. We have to reconstruct, or should I just say guess, at what happened in-between. Nevertheless, it is clear that Feynman does in fact not have the "background" to follow what these students are saying. But one cannot put all the blame on him, given that the "rabbis" are themselves confused, and sometimes simply wrong on a number of matters. A few Feynman-style questions might have clarified the situation, but these never came. Why? Possibly because Feynman believes that he is dealing with knowledgeable "Orthodox rabbis." But the Jewish Theological Seminary is not an Orthodox institution, and the "rabbis" are clearly young men just embarking on their studies. Feynman accepts a number of incorrect statements, such as, that the Talmud has never been translated and that new commentary stopped 500 years ago. We are told that a non-Jew is employed to operate the elevator at the seminary on the Sabbath. Feynman with his "background" does not find that strange. But anyone with a real background would know that this is impossible at an Orthodox institution. These, and other matters, clearly indicate that Feynman did not understand what he was dealing with.

One topic of discussion is the Talmud, which Feynman feels qualified to address because of his "background." And what does he know about the Talmud? He knows that it is a book with very large pages where many diverse topics are discussed in a "medieval kind of reasoning." I read this and I start to get uneasy. The students ask Feynman if they might ask him some questions about science. Feynman is delighted. Science is his favorite topic. And what do they want to know? They want Feynman's opinion whether electricity is fire. The question does not make sense, and Feynman asks for clarification. Well, it seems that this Talmud that they were discussing states that one is prohibited from using fire on the Sabbath, so would the use of electricity be prohibited because it is fire?

Feynman is "shocked." He thought that the students wanted to talk about science; instead they just want to better understand what the Talmud is all about. Feynman is shocked, and I am shocked that he is shocked. Where is the surprise that theology students are interested in theology and not in physics? My unease turns to serious concern. And I don't have to wait long before my fears are realized. Feynman explains that electricity is not a fire, it is not a chemical process. Surely the students would understand that; after all, he did not talk about the plasma state of matter that we perceive as flame. The students are persistent. What about electric circuits that make sparks, like actuating an elevator? Sparks don't matter, electricity is still not fire. Besides, you can put a spark suppressor on the electrical contact to eliminate the sparks. But the students are still not satisfied, and Feynman cannot figure out why. He ascribes this behavior to the medieval Talmud and the resulting lack of logical reasoning of the students.

Feynman decides that he needs to straighten out the students' confused thinking and carefully prepares a set of logical traps that will force the students to acknowledge their mistakes. They meet and Feynman springs his trap. But it does not work. Somehow, in a manner that Feynman does not understand, the students manage to evade his trap. Feynman is not amused. He writes the students off as medieval and superstitious, and he goes on his way. But I cannot write the students off. I know that basically everything they said is true. Yes, they are a bit ignorant in certain areas. But this has nothing to do with using electricity on the Sabbath. I don't understand where this discussion went wrong. It must be some sort of rare aberration, because genius Feynman can understand everything. Clearly it is my duty to correct the situation.

I immediately start planning my letter to Richard Feynman. I can just visualize the results. Feynman will read my letter, and the proverbial light bulb will click. He will immediately see the error of his ways and want to make amends. Feynman will abandon physics, join a Yeshiva (school for Torah study), grow a long beard and spend the rest of his days learning Torah. This is an important project, and I devote time and energy to it.

I carefully explain why "fire" does not mean literally the change of matter which we call burning or fire. The students were talking about the heading, or name, of one of thirty-nine categories of activities or work prohibited on the Sabbath. These are collectively known as *melochah*, or simply "work" in English translation. The category of "fire" includes human endeavor to cause combustion and make light and heat. Besides, fire is not really the key indicator of this prohibited activity. This *melochah* is known as *mavier* in Hebrew, which means "kindling."

The focus is on the actuation and intention of starting a fire, and not the fire itself. Electricity used to start a combustion fire by use of an electrical striker for a gas stove clearly fits the category of *mavier*. Use of electricity to produce fire-like results, such as light and heat, will also usually fall into the *mavier* category. But certainly not all uses of electricity involve the "kindling of fire." Nevertheless, the actuation of electrical devices is generally prohibited on the Sabbath because most of these activities involve some sort of *melochah*, though only one of these is *mavier*. The Talmud reasoning is not confused and illogical. On the contrary, it agrees with modern science, because Orthodox Torah authorities do not suggest that electricity itself is a form of fire. Rather, electricity can be used to kindle (*mavier*) a fire, or produce a fire-like result such as heat or light. I go on and on trying to straighten out the obviously wrong assumptions that Feynman has made. Finally I am ready. I send off my brilliant letter. This is absolute genius, and I know that Feynman will not be able to resist my arguments.

Alas, having reread my letter as I write these words, I find it more than a bit short of the perfection that I remembered. Furthermore, Richard Feynman never had a chance to see my letter, as he was already incapacitated from the illness that soon would take his life. Still, it was not a waste of effort for me. The experience forced me to seriously analyze what had transpired between Feynman and the students, leading eventually to my examination of the five modes of interaction between science and Torah, which I will discuss below.

There are numerous references for those interested in delving deeper into this subject. I do not know which Talmudic statement the students were talking about, but it is likely the statement by Rav Nassan on page 70a of Tractate *Shabbos*, where he quotes the prohibition about kindling fire found in Exodus 35:3. For a nice, clear, overview I recommend the small monograph, *The Sabbath*, by Dayan Dr. Grunfeld. The author defines *melochah* as "an act that shows man's mastery over the world by the constructive exercise of his intelligence and skill." One can see how the use of electrical devices fits this definition, though certainly not all involve some aspect of fire. On the other side of the spectrum in detail would be the four-volume treatment by Rabbi Dovid Ribiat on *The 39 Melochos*. Issues related to electricity and *mavier* take up more than fifty pages, from page 1195 through page 1256. This is not the simple topic that Feynman thought it was.

Analysis of a conversation

The best way that I can explain the interaction between Feynman and the students is by way of the Abbot and Costello comedy routine that begins with the words, "Who's on first?". Here is a summary for those too young to have been exposed to this hilarious conversation. Two people are talking about a baseball game. One wants to know what is happening, and he asks questions. The other one knows what is happening, and he gives answers. The humor comes from our understanding of their miscommunication. The first person speaks in ordinary English, where the word "who" signifies an interrogation. The respondent provides factual answers, giving the names of the various baseball players. But the players have unusual names. The second baseman is named "Who." This is the name of a person, a proper noun and not an interrogation.

Likewise the name of the first baseman is the phrase, "I Don't Know." And so on. Here is the beginning of the conversation. First speaker – "Who is on first?" Response – "I Don't Know, Who is on second." The real thing is much better than my explanation, and it is very funny.

So what does the above have do with the discussion between Professor Feynman and the students? It seems to me that both conversations are of the same variety, where a seeming common language is not common at all. The only difference is that one makes me laugh, while the other makes me want to cry. The students say, *work.* But they don't mean work. They mean *melochah*, a highly technical, legal term on which I have half a dozen books in my own very inadequate library. Feynman hears, *work.* He certainly knows what work means in physics. He is also smart enough and sophisticated enough to understand that maybe these people are not referring to a physics concept but to the everyday, common meaning. Feynman is confident that he understands what he is told; after all, he has the "background." But he is wrong. *Melochah* is not defined, or measured or explained in units of force times distance, as in physics. Nor does it necessarily refer to a job for which one is paid. *Melochah* is related to these concepts, but it is not the same thing. The students are likewise confused. They know that "fire" is not necessarily a physical fire, but the heading of a category of phenomena within *melochah.* I would hope that they knew that the real thing is actually *mavier*, and not fire. But that would have made no difference. I knew about *mavier* when I wrote my letter to Feynman. But I also talked about fire in the letter in order to keep it simple.

The students know that *melochah* is not really the same as the ordinary concept called work. But this is the usual translation. Besides, it's such a simple concept. Any child can understand it. Indeed, my five-year-old granddaughter already seems to understand more about this matter than Feynman apparently understood. The idea is "obvious" when you grow up with it. There are thirty-nine primary categories of forbidden work (sorry, I mean *melochah*). Use of fire is in one of these categories. Electricity is also in this category, among others. It is obvious that there is a connection between fire and electricity. So it does not seem unreasonable that the students wanted to know what, if any, such connection exists in physics. But the relationship between fire, work, *melochah*, *mavier* and Sabbath is complicated and cannot be reduced to the simplistic question whether electricity is fire. Nevertheless, that is what the students ask. And the genius Feynman takes them at their word. He apparently does not understand what they are really asking him. But he thinks that he does understand. And the students likewise think he understands. This creates a false sense of comfort. The students think that they are using language that the other fellow understands. But he does not understand. *Melochah is melochah*, period. You either absorb the idea as a child or you study, not just read but study, some books. You are not going to get it from a short conversation while waiting for the elevator. Not only *melochah*, but many other matters as well were equally misunderstood in this conversation. It is a classic case of "who's on first." Too bad I could not find it funny.

Once started it is difficult to stop. From the study of what others, wiser than I, had to say, and from the analysis of incidents and stories like the one above, I concluded that there are basically five primary modes of interaction between science and Torah. There can also be a sixth mode if you like, in which the two parties understand that they are unable to communicate and do not try. But I will ignore this one and proceed to the communication modes where the two people, or groups, do attempt to communicate.

Interaction modes between science and Torah

The first mode of communication, or perhaps I should say miscommunication, was described above. The two parties think that they are explaining matters to each other and that they understand each other, even though they may disagree. Unfortunately, it's a case of "who's on first." The more they try, the more they misunderstand. Paradoxically, this mode seems to be most likely when the two parties are specialized in their own fields. For this introduces the use of specialized jargon that has "obvious" meaning to one side but a totally different meaning to the other side. Experts in their field, especially those who acquire the knowledge as children, frequently take many presuppositions for granted. They are used to being understood by those in their own group with whom they usually interact, and they assume that they are understood by others. People of lesser experience and knowledge have more doubts about their grasp of the topic and more difficulty in expressing their position. Hence they will take more care to probe for understanding when explaining matters to others. I am convinced that a discussion with someone of less specialized science experience than a Richard Feynman would have been more productive.

Suppose the two groups do communicate effectively. They do understand each other. But they disagree. One says six thousand years, and the other says fifteen billion.

One says instantaneous creation, and the other says gradual evolution. What do we do now? We can agree to disagree. The Torah believer will say – I believe that Torah is truth, so the science must be false. Now the debate shifts, as the goal of the debaters shifts. Now, instead of trying to convince the opposition to abandon its position, each side tries to persuade the other to accept that it has a sound and rational basis for its position, even for someone who disagrees with it. In other words, the goal is to disagree in a manner where we maintain mutual respect. We would like to avoid the case where Feynman called the students medieval and superstitious. Torah-based arguments will not help here because the scientist does not believe in the Torah. Hence we need to use arguments based on science.

We could argue as follows. I have great respect for science. But when push comes to shove, I am not going to abandon the Torah. After all, we know that science can be wrong, has been wrong and is likely wrong now in various matters. We are not talking here about temporary hypotheses or theories that had to be abandoned as new evidence came in. That is part of the scientific method. There are always more ideas and conjectures than can possibly be true. We are talking here about hard-core science that meets the most stringent tests of scientific process-based scrutiny. Take Ptolemaic astronomy, for instance. It withstood the tests of scientific process for hundreds and hundreds of years. Everything worked as required. Eclipses could be accurately predicted, and an overall framework for the workings of the earth and heavenly bodies was provided. Yet it was wrong. The sun does not revolve around a central earth. You say that science is more sophisticated now. True, but the principle is still the same.

Is there anything of greater impact in recent times than Newtonian mechanics? Yet, when it comes to the workings of the universe as a whole, we need Relativity theory. And Relativity theory is in much better agreement with Torah than Newton's mechanics could possibly be. We read the content of debates between science and Bible in Newton's time and see how many of the problems disappear in light of modern science. I say that your sophisticated and modern science of today will be the primitive science of prior years a hundred years hence. Look at history and see how science and Torah have been slowly getting into closer agreement as time moves on. This is not because Torah has changed, but rather because science has changed. I am not being medieval and superstitious when I express faith in the scientific method and the future of science, which I believe will eventually come into full agreement with Torah.

Conceivably, the scientist hears this argument and, just as I imagined with Feynman, embraces a life of Torah, and we live happily thereafter. But I would not hold my breath while waiting for this miracle. Instead, I conjecture that the scientist would respond something like this [Δ]. You are right. The differences between science and Torah are irreconcilable. I have to pick one, and that is obviously science. How can it be otherwise, given what you just said? Yes, I agree that science has been wrong in the past and is not fully correct now. But you yourself made a point of the fact that science is continually getting closer and closer to the real truth. Einstein is a greater authority than Newton, and Newton is a greater authority than Archimedes. That is not the case with Torah. Is it not a fact that current authorities defer to previous ones, and previous one to yet earlier ones, all the way to Moses? Don't you claim that Moses was the greatest prophet or teacher or whatever, that ever was or will be? This is all I need to make my case. Even if there was somebody called Moses. Which is not a given. Even if he did go up and sit on a mountain for forty days and nights. Would you like to try that, and return sane? Assuming it is all true. You still have the problem of loss of information over time. You get further and further away from the source of your information, which is Moses. But my science-based information gets more and more accurate as time goes on. If I have to pick one, it is clear to me that science has a better probability of getting it right today than your Torah does after all these years since Moses.

A fourth way to discuss science and Torah is to recognize that the original creation took place outside the so-called "natural" rules of behavior of the world as we now know it. Even science tells us that the laws of physics, as we know these today, were not in effect in the initial stages of the big bang start of the universe. This avoids conflict by de-coupling the connection between science and Torah. We argue that the modes of logic of the two are so different that we cannot deal with them together. Here is a simple-minded analogy. I ask you – what color is Tuesday? The response is, I don't know. The answer lets me ask whether Tuesday might be purple. The respondent says, no, Tuesday is not purple. It is true that Tuesday is not purple. But the answer is misleading because of the implication that Tuesday might be some other color. In fact, the whole mode of conversation is wrong starting with the first answer. "I don't know" is a misleading answer because it fails to point out that asking for the color of Tuesday is not an acceptable question in the first place. The logic of color, whether it be wavelength or rainbow or primary colors or the physiology of vision, is not connected to the logic of days of the week or lunar month or solar year. The appropriate response is to note that there is no answer to this sort of question.

Let us now take a version of an often-used example from science and Torah. We are told in the Torah that trees came into being on the third day of creation. Three days later, on the sixth day, Adam cuts down one of the trees. It is now six thousand years later, and modern scientists find this tree in petrified form. An examination shows three hundred rings, and the scientists declare that the tree was three hundred years old when it was cut down. But suppose Adam's behavior was recorded in the Bible and I declare that the scientists are wrong, the tree was only three days old when it was cut down. Is there any useful purpose in debating this matter when time is being defined in different ways? I am not a philosopher, so I am not going to start anything that I cannot finish. But it seems clear to me from this example that there simply isn't any way to falsify either the science or the Torah. They are, in a certain sense, both correct. Surely the scientific evidence will not convince the person who follows the Bible. Anybody who believes that a tree can be created also believes that a tree with three hundred rings can be created. And yet the tree meets all the scientific tests and criteria of a three-hundred-year-old tree.

We come finally to the fifth mode of interaction. This is the one upon which all of the essays in this book are based. It is a synthesis that grows out of the tension of the second and third modes of interaction. I argued that Torah represents absolute, revealed truth. Torah cannot be wrong, by prior postulate. I also argued that science provides discovered truth which is continually in process of moving towards an absolute level. If that is so, then ultimately science and Torah should agree with each other. Is that time now? In some ways it seems very close. Not in all things, but the overall level of agreement is truly astonishing. And the exciting part, for someone who works on the basis of this mode of interaction, is that as the rate of scientific discovery accelerates, so does the level of congruence between science and Torah.

But we have to be careful with our language. When I say science, I mean the most advanced levels of scientific understanding that we have today as exemplified by Relativity, Quantum Mechanics and Big Bang Cosmology. Most people have heard of these fields and have some familiarity with the ideas. But science does not stand still. Even Relativity, which changed comparatively little since its inception early in the twentieth century, is not the same today as it was then. We need to look at the latest ideas available, because that is how we get closest to the ultimate scientific truth. Torah by comparison is ancient. Most commentators involved lived many centuries ago. That is good because we can make a fair comparison without concern that the Torah analysis was somehow influenced to fit the latest scientific ideas. Unlike science, though, ancient Torah does not mean simple or unsophisticated. On the contrary. The ideas in the Torah call for every bit as much, or more, care in analysis than the most advanced science. Most people are not familiar with many of these Torah ideas, just as is the case with the most advanced ideas of modern science. These Torah truths were not much discussed and were maintained on a private basis till fairly recent times. Why someone of my relatively low standing in Torah scholarship should be privileged to have access to these ancient treasures is a question outside the parameters of this essay. Perhaps it is that ordinary mortals are now ready for these truths, having been prepared to accept the unbelievable by the discoveries of science.[ΔΔ] Whatever the reasons, the science is here and the Torah is here. All we need to do is look.

[Δ] As mentioned in the Foreword, this is not a disinterested inquiry to discover the truth. I have already made up my mind, and all I am trying to do is to show why it is not unreasonable to believe as I do. So I am not cheating if I take this aside to correct the common misunderstanding about the process of Torah inquiry expressed by my scientific colleague. It is true that our respect for the opinions, devotion, insight and saintliness of prior generations is as expressed by the scientist. But that does not mean that later generations, who build upon the teaching of prior generations, cannot gain new understanding. Our understanding of Torah is not deteriorating. As an example, consider the following story. We are told that Moses was given the merit to move forward in time and observe a Torah lecture presented by Rabbi Akiva. The more Rabbi Akiva explained, the more depressed Moses became, because he could not understand what this great Sage was talking about. Finally, Rabbi Akiva came to the end, stating that he had thus demonstrated the principles of Torah that our Prophet Moses had received and taught. How is it possible that Moses did not understand this teaching which was presented in his name? He was a prophet and received the teaching directly from the Source. He had no need to explain principles and procedures. He just knew. But Rabbi Akiva was not a prophet. He did not know in the same direct way that Moses did, and is thus lesser than Moses. Instead, he had to reason things out, basing his reasoning on truths given to us by Moses. And, paradoxically, because of this need to reason things through, he was able to explain matters in a manner that Moses did not understand.

[ΔΔ] My rabbi, Rabbi Leonard Oppenheimer (see ***Acknowledgements*** section), pointed out to me "a well known prediction of the Zohar" that "The Gates of Wisdom" will be opened around the year 5600 (1840) "and pour forth in an unprecedented manner ... revealing to mankind the secrets of hidden knowledge." I tried not to dwell on the implications that I had not previously known about this "well known prediction." Instead, I tried, in vain, to find some major scientific discovery around 1840 that would correlate with an increase of wisdom and secret knowledge. True, nothing on the level of Newton's work on gravitation or the development of quantum mechanics, occurred around 1840. But a number of lesser, but cumulatively critical scientific developments, including the work of John Dalton on gasses and chemistry, Michael Faraday on electric machines, Andre-Marie Ampere's investigations into the nature of electricity, and the insights into the connection between electricity and magnetism by Hans Christian Oersted, all came during this period starting in 1820. And then I read an opinion piece in the October 16, 2000, issue of *Newsweek* magazine, where commentator George Will, quoting DeMuth, assures us that 1820 "marks the beginning of the modern age" where "science" has displaced other drivers of "social and economic development." The year 5600 (1840) is a nice round number counted in centuries, and I am happy to accept 5580 (1820) as close enough. Possibly the Zohar explains why as well as when the hidden knowledge will be revealed. But I am not aware of that. All I can say is that a Torah-based source indicates that ordinary people will be privy to the kind of matters I am discussing in this book.

Chapter 2

What Answer Would You Like?

A Word About Probability

We know with absolute certainty that we have our kind of universe. But why this universe and not some other kind? Indeed, why do we have any kind of universe in the first place? What are the odds of getting our kind of universe? The answer calls for a recourse to probability, and probability is a tricky field. Permit me to illustrate with a little story.

Three people applied for an accounting job. The interview consisted of one question – how much is two plus two? The applicant with the marketing background asked for a one-week time extension so he could convene a focus group. The applicant with the engineering background pulled out a calculator and announced that two plus two equals four to eight significant places. She would need a more powerful calculator to provide a more accurate answer. The applicant with a background in statistics asked – what answer would you like? A mildly amusing story. But in a certain sense, it is true. It is well established that while the world is full of lies, it takes statistics and probability mathematics to create really big, or subtle lies. This form of mathematical reasoning can lie to us on, at least three levels.

The simplest level is the lie of incorrect usage or calculation. This field is frequently very much counterintuitive. Relying on intuition can easily lead to a false result. Consider the following well-known puzzle that most people get wrong. A contestant wins the right to choose one from among three sealed envelopes. One of the envelopes contains a fabulous prize, while the other two are empty. The contestant picks one of the envelopes at random, and all would agree that he has a one in three chance to win the prize. This is not just a logical or mathematical deduction. We can test our conclusion by experiment. A one in three chance means that the desired envelope will be chosen one third of the time if we perform a large number of unbiased and identical trials.

Now comes the tricky part. The contestant chooses one envelope and leaves two on the table. Whereupon the game-show host, who knows the contents of each envelope, opens one of the discarded envelopes and shows the contestant that it is empty. The contestant is now given an opportunity to exchange the envelope he chose for the remaining one that he originally did not choose. The question before us is whether there is any advantage to making the exchange. Most people would say, no. All envelopes are of equal statistical value. After all, the alternate envelope could just as well have been chosen the first time around. We would then be considering an exchange of the discarded envelope in favor of the one the contestant already has. We don't know what is inside the envelopes. So they are all of equal value. This, and other such reasoning, seems reasonable. But it is wrong. There is actually quite a significant advantage to switching the envelopes. I will explain the reason below, and you can always run a few trial experiments to see that the opportunity to get the prize is much better if the envelopes are switched.

There are a number of ways to explain why it pays to switch envelopes. Here is my favorite. Suppose you had the choice of picking one envelope, and keeping whatever prize it contains, or picking up two envelopes and keeping whatever prizes these contain? Clearly your probability of getting the good stuff is greater if you choose the two envelopes. And that is precisely what switching after the fact achieves. You discard the one envelope you previously chose, and the two remaining envelopes are opened instead. The game-show host opens one, and you open the other one. It makes no difference who opens these two envelopes as long as you get to keep the prize. In fact, the game-show host could just as well have opened the two remaining envelopes. This clearly offers two out of three chances to win. Switching envelopes provides a significant advantage, but it is not intuitively obvious even though this is a comparatively simple situation with only a few probability states. How much more likely to get into trouble when we deal with large, and I mean very large, aggregates of possibilities?

Another version of getting whatever answer you like, involves lack of clear definition of terms. The job applicants were asked for the answer of two plus two. It is obvious to everybody that the answer is four. But is it? Two centimeters plus two meters is not four. Walking two miles south and then two miles north does not bring you four miles from your starting point. One must exercise great care to avoid falling into this trap.

Finally, we have the so called "circling the arrow" fallacy when dealing with large numbers. This is probably the most important factor in our discussion. Again, a story to illustrate the concept. An archer wants to demonstrate her prowess in marksmanship. She sets up a very large wall with ten million tiny targets on it. She then circles the target that she hits after shooting an arrow from a great distance. The archer then explains, truthfully, that she hit a target that has a one in ten million probability of being hit. The false implication being that this involves exceptionally skilled shooting. But the result implies no such thing. The key point is that the target was circled after the fact. She wasn't aiming at it, so hitting this target is not different from hitting any other target. Now, if the target had been circled before the arrow was released, then the archer would really have something to boast about. We need always to be alert for instances when we are, figuratively speaking, circling the target as we address the impossible odds of getting our kind of universe and our kind of life by pure chance.

A word about miracles

I have never participated in a certified miracle. But I have been involved in something that was legally classified as miraculous. I am talking here about a so called "act of G-d." Colloquially, this classification invariably refers to an undesired situation. Apparently the good stuff qualify as acts of humans. As they say – grab the credit and pass the blame. Anyway, it happened that a sizable area of Portland, Oregon, lost electric power in the winter of 1998. This is not unusual. We have power outages several times a year. But this one was different. The wind toppled an ice-laden tree onto a 7.5-kilovolt distribution line, which touched a household electricity distribution line, sending a high-voltage surge into my house. Most electrical appliances, including those that were not turned on at the time, were damaged as a consequence. I soon received a letter from PGE, the local electricity utility, which explained the following. PGE had done nothing wrong. The probability of such a rare cascade of events was too low to plan for. Nevertheless, there were safeguards in place. The safety circuit breaker opened up well within the specified regulatory time limits. The high-voltage spike, along with all electric power, was quickly disconnected from the distribution grid, and no serious damage was done. This was an exceedingly rare occurrence for which no person or human agency was responsible. PGE did not use this exact expression, but what was described is generally known as "an act of G-d." I thought at first that the language was just a way for PGE to shift blame. But no. My insurance company took the same position, and paid for repair or replacement of the damaged items. Yes, this was a legally certified miraculous occurrence. But why?

What is miraculous about snow in the winter? Don't we have that all the time? What is miraculous about ice accumulating on a tree? Is that something new? What is miraculous about high wind gusts in the winter? That is the normal pattern. What is miraculous about soil getting weak after months of steady rain? Surely we understand the science of dry and wet soil. What is miraculous about the roots of an ice-heavy tree breaking loose from rain soaked soil in a heavy wind? The force levels can be calculated. What is miraculous about a tall tree falling onto power lines strung below? What else would you expect? Perhaps the tree should fall up instead of down? What is miraculous about PGE choosing to place the high voltage line where they did? They had to place the stuff somewhere, after all. What is miraculous about the economic cost calculations that caused PGE to run household electric power lines on the same poles as high voltage lines? Do you really think it strange that they want to save on costs? What is miraculous about the high-voltage line falling on top of the household voltage line below? What is miraculous about a 110-volt appliance getting fried when subjected to thousands of volts?

Everything is natural here – ice, snow, wind, rain, trees. It all happens all the time. We are used to it. All the actions are human actions here – proximity of power lines to each other and the tree, not trimming the tall tree, etc. People made these choices; the "accident" did not happen by itself. Where is the miracle? There is a simple answer for those who accept the Torah. Everything that we call natural is in fact a miracle. The universe is a miracle, and all it contains, including ourselves, is miraculous. It does not have to be this way, but it is. It is all this way because it was deliberately chosen to be this way. Furthermore, it was not only chosen to be this way, but is continually maintained this way. Nothing could or would exist to start with, nor could or would it continue without outside intervention. And that makes it miraculous indeed. That is what the Torah says. But that is not, I think, what PGE would have meant by an act of G-d, had they actually used this phrase.

PGE was talking about a result that is so unlikely that no one could be expected to anticipate it. Furthermore, we have many such factors. And most of these, such as ice and wind, not only cannot be predicted, but they cannot be prevented. Clearly one of the indicators leading to a designation as miraculous is that we have a cascade of many unusual events. No one who does not, *a priori*, accept Torah would designate rain as miraculous in

Oregon where it rains most of the year. How rare is the combination of factors and results that I described in the previous story? I do not know. But it is certainly not common. The front page newspaper story did not mention any previous such occurrence. I could try for an estimate based on the assignment of probability levels to each of the separate factors that together constitute this event. Thus: rain in the winter is a 90% certainty, and snow and ice in any winter is about one in three winters. How many power poles in Portland carry high-voltage and household voltage lines side by side? How many such lines are beneath tall trees? How likely is it that the tall tree is not trimmed by PGE for safety? How likely that any one tree will topple in the winter rains and snow and ice? And so forth. I do not know what the final result would be. But I can conceive of a result of less than one chance in ten thousand or even one hundred thousand that this cascade of events would occur for a particular tree at a particular day. But we don't really care if it happened on one day or another day, and for this tree or some other tree. This increases the probability significantly given the very large number of trees that can topple throughout a typical winter. But let me instead be more generous, and make it one chance in one million, or in scientific notation as ten to the sixth power (10^6). This is big odds. But we can easily relate to this number. We are used to people winning lotteries against even bigger odds. But the probability of getting the kind of universe we have, with the kind of creatures that we have, is in a different category altogether. There just isn't any comparison. Furthermore, it isn't just one single number or event that we need to deal with. Just as with the electric power, we have a cascade of incredibly impossible events to contend with. We need to look at these numbers to see if we can learn something here. But before we do I would like to go back for a bit more exploration of our notions about miracles.

We live in a stable universe with predictable rules and results. The position of Torah is that predictability and stability are necessary so that human beings can exercise their free will, and be responsible for their actions. "Natural" phenomena that fall far outside of normal patterns are few and far between. And even then, a unique event usually has the possibility of being explained as a seldom occurring part of the regular pattern. Thus when the people complained that they had no meat in the desert, "A wind went forth from Hashem and blew quail from the sea..." (Num. 11-31). Can we not explain this as an unusual weather pattern which is nevertheless a normal part of the so-called "natural" world? And even if the "natural" mechanism is not as obvious as in the above case involving wind, will not the scientist postulate a natural law theory to explain the anomaly? Our Torah sages tell us that we should not concentrate our attention on nature-defying miracles. Rather we should instead look at that which we call natural, because therein lies the greater miracle.

Here is how Rabbi Eliyahu Dessler explains it in the second volume of *Strive for Truth*, the English language translation of his *Michtav Me-Eliyahu*: "On rare occasions, because of some special need, and for someone who has extraordinary merits, Hashem will override the laws He has written into the cosmos and perform open miracles which have no physical cause." Those who accept Torah will accept that there was a miracle. But those who do not accept Torah will not accept the validity of even an open miracle as defined above. Here is how Rabbi Dessler explains it a few paragraphs later: "They would look for, and find, some twist in a physical theory which would account for the miracle in materialistic terms, or they would invent an entirely new theory; in one way or another they would inevitably assimilate the miracle to the naturalist scheme. The miraculous would become 'nature.'" The non-believing scientist will usually not accept the idea of an open miracle. Furthermore, the rare occurrence is not conducive to systematic investigation. But that does not mean that we cannot examine miracles, because "nature" itself is the greatest miracle. Hence we need to examine the laws of nature to look for what might be identified as a miraculous pattern. Before proceeding to this, though, a few more quotations from Rabbi Dessler about nature as miracle.

"What is the difference between the natural and the miraculous? Are they not both from Hashem? ... We call G-d's act a 'miracle' when He wills an occurrence which is novel and unfamiliar to us and which consequently makes us aware of the hand of G-d. We call G-d's acts 'nature' when He wills that certain events should occur in a recognizable pattern with which we become familiar." An important factor of "nature" is that it is not self-maintaining. We are taught that it has to be maintained, indeed, recreated by G-d moment by moment. The clock was not wound up and abandoned. Torah tells us that G-d is involved in the world in an ongoing and positive manner. Rabbi Dessler comments on the twice-repeated blessing that we say in the morning, "He Who renews each day continually the works of creation." He explains that this is to "teach us that we are not to consider the physical creation as something created once at a point in time in the past and since then existing by its own power, but as something that needs to be 're-created' each day and each moment – 'continually'. It follows that the coming into being of each successive moment is a new act of G-d's will … ."

About big numbers

We can now examine the scientific evidence and relate it to the Torah position about miracles. But first we need to internalize the implications of big – really, really big – numbers. The largest named regular numbers, not in the infinity range, that I know about are googol and googolplex. These are a one followed by 100 zeros, or ten to the one hundredth power (10^{100}) in scientific notation, and a one followed by a googol of zeros which is ten to the tenth power which in turn is raised to the hundredth power ($10^{10^{100}}$), respectively. I don't really know what sort of examples or analogies to connect to these numbers in order to create a feeling of what we are dealing with. But I shall try, because some of the probability odds we need to discuss start to approach a googol and in one, incredible, case the result is even greater than a googolplex.

Let's start with some distance comparisons. The best current understanding on the age of the universe is 14 or 15 billion years ($15\text{x}10^9$). This means that the distance in the universe that we can observe is 15 billion light years.[Δ] A light year is the distance that light travels in one year and is 9.46 million million kilometers, that is $9.46\text{x}10^{12}$ km. Using these numbers, we find that the extent of the universe is $1.42\text{x}10^{28}$ centimeters. The electron is the smallest particle, excluding sub-atomic or exotic particles. The radius of the classical electron is $2.82\text{x}10^{-13}$ cm. Dividing the distance of the universe by the radius of the electron, we find that we can lay $5\text{x}10^{40}$ electrons end-to-end across the universe. This is a nice number to have, because some of the probabilities we shall look at are in this neighborhood.

We've dealt with space. Let's take a look at time. The universe is 15 billion years old. This translates to 4.7×10^{17} seconds. This is a huge, huge number. But not impressive on the scale that we've been dealing. But we can do better. A second is an arbitrary, human devised unit of measure. The smallest "natural" time unit is 10^{-43} seconds, known as the Planck time. Hence the universe is 4.7×10^{60} Planck time units old. This is very nice, because 10^{60} is a rather well-known, and often quoted, probability value in big bang cosmology. Now we have something to compare to. But still no googol. How about moving forward in time? That seems to be pretty much open to us. But not really. Because eventually the universe will deteriorate and pretty much come to an end. There are any number of end-of-universe calculations, depending on the choice of assumptions. Certain common assumptions lead to a final age about one hundred thousand times the current age. This adds a puny 10^5 multiplier. Looks like time will not get us to a googol.

We tried space and time. How about density of radiation photons over the whole volume of the universe? It's a big universe, and there is lots of radiation. The number of radiation photons per unit volume depends on the temperature. One can calculate that the whole universe is bathed by about 413 photons per cubic centimeter for the remnant of the big bang start of the universe, the current 2.73 degrees temperature cosmic background radiation (CBR). The radius of the universe is 1.42×10^{28} cm, as given in the previous paragraph. Taking the observable universe as a sphere of such radius, the volume of the universe is 1.2×10^{85} cubic centimeters. Hence the total CBR quanta from the big bang is 5×10^{87} radiation quanta. Think about it for a while. This is the total stuff from the big bang, the beginning itself. This is the whole universe. Yet a googol is more than that.

How to get to a googol? Apparently it takes some really crazy stuff to do it, if the above is not sufficiently bizarre. I've taken the volume of the universe, previously determined to be 1.2×10^{85} cubic centimeters and divided by the volume of the classical electron, at 9.4×10^{-38} cubic centimeters. This finally gives a result greater than a googol, at 1.3×10^{123}. Yes, we know that we cannot talk about the volume of electrons in this way. Indeed, talking about the volume of the universe this way is not really proper either. And we know that we cannot pack spheres inside spheres without loss of empty volume. And electrons cannot get this close to each other anyway. As I said, it takes some pretty crazy assumptions to come up with some sort of physical analogue of a googol. That is the whole point. A googol is an amazingly large, an impossibly large number. Yet one of the improbable probability numbers we shall discuss in the next chapter is a googolplex. What sort of odds does it take to declare a miracle?

Δ The event horizon, also known as the horizon distance, is the distance that a light-speed signal can reach during the existence of the universe. This would be the age of the universe in light years, for a static universe. But our universe is not static. It is expanding. And as the fabric of space stretches so does the signal transmission. As a consequence, the horizon distance is much greater than the distance that a light signal could reach in the roughly 15 billion years that our universe has been around. It can get a bit complicated to determine the horizon distance. Guth estimates a distance of about 30 billion light years in his book *The Inflationary Universe*. I have chosen to make my life easy by using the static universe value of 15 billion years in this essay and in others, as well. This is definitely way too low a value, and scientifically speaking, incorrect. But the point I want to make works just as well as with any other event horizon value. And it keeps me out of side issues that I would like to avoid.

Chapter 3

So What, And Who Cares?

On Miracles And Nature

One of my practical minded professors in school liked to keep us attuned to reality in the midst of theoretical discussions. You knew that you had gone too far into theoretical speculation when he asked, "So what, and who cares?" I discussed in Chapter 2 some ideas about miracles, and provided a number of strange examples involving probability calculations. It is time now to answer these two important questions – so what, and who cares? What practical conclusions can we draw from these speculations?

About miracles

The last sentence in the previous chapter asks – what sort of odds does it take to declare a miracle? Different people have different ideas, and the answer might depend on what we mean by "miracle." Here I would like to consider three categories: suspension of natural law, continual maintenance of natural law, and the initial creation.

We begin with the usual, everyday meaning of the word miracle. An important feature is that this is a unique, an extraordinary, event that transcends the understood natural order of the universe. Such a happening is ascribed to Divine intervention and termed miraculous. It is an "open miracle" in the words of Rabbi Dessler in Chapter 2. Attempts have been made to find current physical residual evidence of such miracles in the past. This author does not feel qualified to attempt this. There are two basic reasons. First, Torah Sages tell us that an open miracle can produce only a temporary result. Hence it would not be unusual that "no permanent record is left of miracles" as explained by Rabbi Aryeh Kaplan on page 39 of the book *Immortality, Resurrection and the Age of the Universe*. Furthermore, even if a permanent record is left, it is most likely that science will find a way to incorporate even the uniquely miraculous into the overall natural framework, as explained in Chapter 2 by Rabbi Dessler.

Consider the miracle of the parting of the sea. We are told in Exodus 14:21 that, "Moses stretched out his hand over the sea; and the Eternal caused the sea to go back by a strong east wind all the night." Nachmanides explains that "it was His will, may He be blessed, to divide the sea … [only] making it appear as if the wind dried the sea." The natural seeming event "caused the Egyptians to err …, they thought that perhaps it was the wind which made the sea dry land, but that it was not the power of G-d." Consequently the pursuing army thought that it was safe to follow. But they were wrong. This was a miracle wrought by the power of G-d through "natural" seeming means. It was not safe to follow. In retrospect, given subsequent events, it is clear that such a happening would have to have been a miracle. But people today also agree that the effect was temporary, without any remaining residual evidence that we can examine.

Just a few verses later, in 14:31, we have a reference to "the great hand of G-d." Nachmanides explains that this phrase has important Kabbalistic implications, and he learns various things from this. "It is by this attribute [which the "hand" points to from Kabbalistic teaching] that the sea was divided." In other words, the "hand" is connected to miracles. The reader will have to wait for the analogy of a "hand" in connection with the science of the Higgs field and resulting false vacuum in Chapter 11, to see how science uses a similar analogy in connection with results that we might call miraculous.

We face a different set of issues when explaining the well-known story of the flood. For years Biblical critics claimed that it was a work of fiction and there was no factual basis for it, due to lack of physical evidence. Then in 1999, Ryan and Pitman published their book, *Noah's Flood*, suggesting a "natural" mechanism for such an event. The following year, on September 13 of 2000, *The Washington Post* published a report titled "Relics of Noah's Flood?" The story describes how archaeologists, using modern technologies, such as side-looking radar and robotic submersibles, found man-made structures at the bottom of the Black Sea. The waters at the lower levels of the Black Sea are anoxic, that is lacking in oxygen, which provides excellent preservation of organic materials over very long time periods. Analysis of these discoveries shows that a wide-ranging and powerful flood wiped out all life in the area thousands of years ago.

This physical evidence confirms that the Bible provides a most accurate historical record, in this case as in so many other instances. But does this confirm and "prove" the miracle of Noah's Flood? No, say the critics. It just shows a basis for the "legend" told by the scribes who composed the Bible. You want to "prove" a miracle? Then tell me how the animals from the various continents got to the ark? And how the ark was able to accommodate so many animals? And how … a few more times. My answer is that I do not know, precisely because it was a miracle. And there we stand. Those who believe, believe, and those who do not, do not. There are those, more brave and more knowledgeable than I, who are willing to deal with the fine details of this miracle. But this writer is not among this group. Rather, I assign this area to the science-Torah communication mode where the two do not interact, as described in Chapter 1. Miracles are miracles, and science is science. One cannot falsify the other, and we are free to believe in both or in neither, as we wish. I will not attempt to use this type of miracle to draw any science-based inferences about a Creator.

The miracle of existence

The Orthodox Jew believes that there is only One reality. All else was created and fashioned by this One. As we read in Isaiah 45:7, "… [K]now that there is nothing besides Me, … I am the One Who forms light and creates darkness …." Light can be formed, that is, changed as to its properties or composition. Furthermore, we know from science that light, or to use the more correct scientific term, energy, is equivalent to matter, and one can be transmuted to the other. But not so with darkness. This is not ordinary darkness which is the mere absence of light. Rather this is primeval darkness which is a something created out of nothing; the highest form of creation. We have here the same Hebrew word, *barah*, as in the very first sentence of the Bible, referring to the creation of initial time and space. Such a substance, if one may use the word substance here, cannot be formed or fashioned. One can form or fashion a substance that already exists. But this darkness must first be brought into existence. It is created.

Not only is all initially created and then adjusted and manipulated to fashion the universe we now have, but this universe does not just run or operate by itself. Rather it is maintained in operation by the One who initially brought it into being. The creation and establishment of the universe and maintenance of the universe go together. Thus, we have the following statements within a few words of each other in the morning prayers: "Who forms light and creates darkness … in His goodness renews daily, perpetually, the work of creation." Psalm 7 talks about the universe which is ours to enjoy. "When I behold Your heavens, the work of Your fingers, the moon and the stars, which You have set in place …." Referring to the four letter Name in connection with this psalm, Malbim (19th century commentator) explains that this, *Shem Havayah*, is the name of *perpetually bringing forth into existence*. Not only matter and energy, but as Rabbi Munk tells us on page 86 of *The Wisdom in the Hebrew Alphabet*, "Even the seemingly irreversible laws of nature … are continually renewed." We also see the same maintenance concept in the quotations from Rabbi Dessler in Chapter 2.

There are two types of miraculous results here. One involves the maintenance of the universe in operation, and the other involves the initial creation and fashioning of all that exists in the universe, including ourselves. Let's start with the perpetual care and maintenance of the creation.

Is there any scientific evidence that the universe is being continually renewed and perpetually maintained? Yes, there is. But we have to be careful how we use this evidence. The mere fact that the universe might be in a state of continual renewal or maintenance does not prove that this is directed from an outside Source. The scientific, natural, laws of the universe, as we understand these today, account for the renewal results. Hence, we cannot use this information as scientific proof that a Creator exists. But we can use this discussion to argue that science and Torah are compatible, and that science does not falsify Torah. We can also argue that the fact that ancient Torah fits current science in this highly esoteric area says something about the probability that there is more than mere "coincidence" involved here. But statistical inference is left to the third type of miracle, so we will not deal with that here. All I want to illustrate here is that modern science can justify the statement about continual renewal of creation.

Ongoing creation

Many of the most advanced modern electronic devices owe their behavior to quantum mechanical effects. To say that quantum mechanical behavior is counterintuitive is an enormous understatement. No less an authority than Albert Einstein refused to accept that the universe works this way. I am an electronics engineer and have had occasion to use such devices. There are instances that require a device that can change, or switch, its state very rapidly in order to make a switch that has a very fast on-to-off time transition. But electrons take time to move from one place to another, especially when they need to overcome an intervening barrier that separates different parts of a device. Here is where we can take advantage of quantum mechanical sleight-of-hand type behavior. We might use a device called a "tunnel diode" that exhibits very fast switching time. This is because an electron does not transition across the barrier. Rather, the electron is said to "tunnel" through the barrier due to quantum mechanical effects. Essentially, the electron disappears from one side of the barrier, and almost instantaneously an identical electron reappears on the other side of the barrier. Indeed, quantum mechanics is full of disappearing and reappearing fundamental particles. Even more startling is the idea that new stuff can appear out of a so called "false vacuum." The whole universe might have come out of that, and we will discuss this in more detail when dealing with quantum mechanics. The essential point is that we have a physical principle for the disappearance and reappearance of fundamental particles a short time later.

Of importance to our specific discussion here, is that according to quantum mechanics, fundamental particles like the electron are made to disappear, and instantaneously an identical version reappears. This happens all the time, everywhere. In effect, the universe is being perpetually renewed, just as the Torah says. Here is how Rabbi Dessler explained this matter more than fifty years ago. I want to emphasize that Rabbi Dessler is not expressing a personal opinion, but rather explaining in modern terms an ancient Torah-based idea. "When we observe an object which appears to persist in the same form over a period of time, we certainly think that today we see the identical object that we saw yesterday. But it is not so, … Hashem re-creates for us momentary sections of existence …" Current ideas on quantum mechanical effects for the very small point to substantially the same idea. It is true that some of the preceding is still uncertain theory. But some is very much verified by the fact that millions of people are using devices mass-produced along these principles.

Not only is the existing universe being recreated, but new parts of the universe are constantly coming into being. The expansion of the universe is as close to an established fact as we can get in cosmology. But there is nothing outside the universe to expand into. The word "nothing" does not mean empty space without matter or energy. This would be a vacuum, and a vacuum is an empty something and not a nothing. This type of something was created out of absolute nothing in the first phase of the creation of the universe. What many people do not realize is that the initial creation has not stopped. New space is being created continuously as the universe expands, just as the Torah states.

Finally consider the following example of continual renewal and creation. Particles do not possess a real existence in the standard, known as the Copenhagen, interpretation of quantum mechanics. Rather, we have a superposition of quantum states. It's sort of like an addition of waves. The particle only acquires a solid, or real existence, when the so-called "wave function" is collapsed by an observation or measurement. Without an observer we just have spread-out quantum state waves rather than space-limited objects. This is bizarre enough when dealing with the very small, like an electron. But a quantum mechanical understanding can be made to apply also on a larger scale. Indeed, it can be done for the whole universe. Science philosophers have a problem as soon as that happens. We ask the question – who observes the whole universe so that its wave function will collapse into the physical reality that we have? This is not a joke. The standard system works very well in dealing with the world on the small scale. Yet an alternate quantum mechanics mode of thinking, known as the Many Worlds interpretation, competes with the standard model primarily because this avoids having to answer the question I just asked. Need I say Who, in the opinion of this author, maintains the universe in existence by continually and perpetually collapsing its wave function?

The googol numbered target range

We now come to a discussion of nature as miracle. Here we have the opportunity to introduce the idea of a proof on the validity of Torah. But we have to be careful. The proof is statistical, and we know from Chapter 2 how easy it is to go astray here. The question we ask is very simple. What is the likelihood that our kind of universe and our kind of life would come into being without some sort of prior plan or guidance? The answer is not so simple. Not on account of the mathematics, which is relatively straight-forward, but rather on account of the underlying assumptions. Here is where we can play the game of "What answer would you like?" I know of only one way to eliminate all possibility of partisan bias and ensure a high level of integrity of results. I will try as much as possible to use assumptions and statistical results provided by those who do not accept the Bible as a source. This does not prove that the numbers are correct, but it does eliminate the possibility of even unconscious bias in favor of a Torah-based explanation.

We begin with the probability representing the likelihood that a universe capable of supporting life could develop by blind chance. This probability was calculated by the distinguished Oxford mathematician, Roger Penrose, as 10 to the 10th and in turn to the 123rd power. This is one chance for life out of more than a googolplex possibilities. It took, in Chapter 2, the totally impossible analogy of having the whole universe stuffed with electrons, all crushed against each other, to achieve a level of 10 to the 123rd power. Yet here we have a well-known, reputable scientist claiming odds against a universe of life at 10 raised to that power. Perhaps somebody can come up with some sort of physical analogy or example representing such a number, but this writer is fresh out of ideas. The only thing I can say is that this is about as close to zero chance as I can imagine without actually saying zero. Note that calculations about evolutionary gene mutations (a matter addressed in the next chapter), some of which run into the googol range, is not involved here. But this would actually not change anything. What is one googol, more or less, out of a googolplex? Nothing. And that is the whole point. The odds for us being here by chance is so small, that a googol is as nothing by comparison!

Okay, ask the question. So an impossibly large array of different universes is possible. What does that prove? One of them has to be here, and it happens to be the one that we got. Is that not a case of circling the target after you shoot the arrow, as discussed in Chapter 2? No, it is not. That discussion applies to identical targets. But these targets are not identical. The target that was hit happens uniquely to favor life, while the others do not permit life. It really is a case of choosing one out of a googolplex. The target that favors life was circled in advance. You shoot and you hit the circled target. You then do it again, and several times again. Sir Martin Rees, Astronomer Royal and Professor at Cambridge University, tells us in his book *Just Six Numbers* that it took six perfect shots to make our universe. The numbers range from the huge ratio of electrical to gravitational forces for the proton at 10^{36}, to the tiny cosmic-structure indicator at 1/100,000, and values in between. These numbers appear to be unconnected to each other and they come to us out of nowhere. Rees tells us on page 161 of his book that we would not be around, "Had these numbers not been 'well tuned'..." Is it all blind chance, or is the arrow somehow guided to the unique, life-giving, target? Let us examine a few of the many improbable hits on the targets of life.

In the beginning was order and balance

Let's start at the beginning. The beginning is the Planck time of 10^{-43} seconds, the smallest time interval that the peculiarities of quantum mechanics permit us to deal with. The baby universe was expanding at a furious pace. But everything was perfectly balanced for life. There is a critical expansion rate where everything has to be balanced to one part in 10^{60}. Had the initial expansion been weaker, everything would soon fall back together. Had the expansion been stronger, there would be no opportunity to form clumps of galaxies and stars. Inflation theory, which is pretty much accepted as part of the standard big bang model, shows how such an exceedingly well-balanced, early universe comes about. Change this "density parameter" by just one part in 10 to the 60th power, and we would not be around to discuss this. How much of a change are we talking about? You will recall from Chapter 2 that the total age of the universe is just 4.7×10^{60} Planck time units. Chop up the total time of existence into the smallest possible time units, and you get about 10^{60}. The initial behavior of the new universe had to be balanced to that degree.

It is a basic principle of physics that a closed system will transition over time from a state of higher order to a state of lower order. This is expressed as the second law of thermodynamics. Currently, the universe is in a state of substantial order. How did it get that way? Perhaps matter just arranged that way by chance over a long period of time? Ludwig Boltzmann calculated how long this would take. The number is 10 to the 10th power raised to the 80th power, or almost a googolplex of years. Anything is possible if you wait forever. But we cannot use forever in our calculations, because the universe is a mere 10^{10} years old. The alternative is to start out with some uniquely special conditions. Penrose has looked into this and estimates the odds are one in 10 to the 10th to the 30th power against this. So how did we get where we are? The Torah has an answer. First came creation of the universe, and then the matter and energy was shaped and molded into the state of order that we now find. It took six days to do it.

We are waiting for a unified theory that will explain everything through the agency of a single force. Currently, though, physics is expressed by means of four forces. The weak and strong nuclear forces, and electromagnetism, and gravity. Among other things, the balance between gravity and the electromagnetic force determines what kind of stars we have. Stars are held together by gravity and lose internal energy through electromagnetic radiation. Balance these forces one way and you get a hot blue giant star; go the other way and you get a cool red dwarf star. Neither one of these is good for us. We need a star between these two. And that is what we usually get, a medium temperature, main sequence star, like our sun. But not really like our sun, because our sun is larger than 95% of the stars. So while main sequence stars are fairly common, our type of star is actually quite rare.

Furthermore, the balance of forces in the star has to be just right. Thus, the force balance between gravity and electromagnetism has been calculated to have a sensitivity of one part in 10^{40}. Change the balance between these forces by one part in 10^{40} and we would not have our type of sun or our type of life. What sort of odds are these? We calculated in Chapter 2 that if it were possible to place electrons side by side all across the universe, there would be 5×10^{40} of them. This is like saying that we would not be around if just a few electrons moved from one part of the universe to another part of the universe. What does this mean?

What about carbon?

There are more numerical examples. There must be more if we are to reach the googolplex odds discussed in the first example. But perhaps some descriptive illustrations might be easier to relate to. Let's consider the formation of the element carbon, upon which all terrestrial life is built. No carbon means no us.

The currently accepted standard model for the universe posits that everything began with a so called "big bang" about 15 billion years ago. The initial big bang universe was very small, very dense and very hot. It consisted of energetic matter particles mixed with radiation quanta in a hot mixture at equilibrium distribution. New particles of matter constantly sprang up out of $E=mc^2$ type energy-to-matter conversions, and likewise matter was annihilated back into energy. This super small, dense and hot universe expanded very quickly, and it cooled as it expanded. Neutrons and protons started to stick around at a temperature of ten million million (10^{13}) degrees on the absolute Kelvin scale, where the energy of radiation photons equals the rest-mass energy of neutrons and protons, and the environment was cool enough not to turn all of these back into radiation. This is the beginning of the element hydrogen. The energetic high temperature neutrons and protons knocked into each other frequently. And when the collisions were just right, they stayed together. We now have the formation of helium-4, consisting of two protons and two neutrons. In the meantime, the universe was expanding and cooling very quickly. In fact, this process proceeded so quickly that the temperature and density soon dropped below the point where higher level elements could form. The result is that the matter of the early universe consisted of about 74% hydrogen, 26% helium (by weight) and very little of anything else. So where does all the other stuff come from? It was cooked in stars and spread out in supernova explosions.

Let's take a look at what it takes to get a supernova and the implications of our being around to talk about it. A heavy star gets more compressed and hotter as it burns more and more complex elements. Finally, at a temperature of over a billion degrees, silicon is transformed into iron-56, and the process ends. Soon the iron core of the star is compressed with enormous force, as gravity takes over without the counterbalance of the heat and pressure from the burning fuel. The iron turns into a ball of pure neutrons. But neutrons have a very much more compact structure than iron, and the whole core collapses in size. A ball about the size of the sun suddenly collapses to the size of a large mountain. We now have a neutron star core. The surrounding material falls onto the neutron star core at a speed which is an appreciable portion of the speed of light, and the shock wave tries to squeeze the neutron core even further. But the core cannot really be compressed, and it bounces back in an enormous explosion which is the beginning of the supernova process.

But now we have a problem. The reaction from the neutron core is not strong enough to scatter all the surrounding material, including the life-giving carbon, into the cosmos. The reverse shock wave needs a boost. The boost comes from a heavy flood of neutrinos. Neutrinos are very strange particles. They simply don't interact with matter. So how come there is an interaction in this case? Everything works out just right, provided the weak force is of exactly the right value. The weak and strong forces, electromagnetism and gravity are the four basic forces of this universe. Change the weak interaction a tiny bit to one side or the other, and we would not be here. The same weak interaction is also involved in determining the ratio of hydrogen to helium in the early universe. We need both supernovae and a reasonable amount of helium, and the range of weak interaction that permits this is very narrow. So why is the universe this way? I do not know.

Now comes the interesting story of that element of life, carbon. Helium is a very stable element. It is so stable that for a while physicists thought that it was a fundamental particle, and it was named the alpha particle. Carbon has a mass number 12, and consists of three helium atoms stuck together in a stable configuration. Unfortunately, two helium atoms, which make beryllium, are very much unstable. The stuff sticks around for less that 10^{-16} of a second before disintegrating. It takes an additional neutron to make stable beryllium-9. Therein lies our problem. A collision of stable beryllium-9 with helium-4 will not add up to carbon-12. The unstable beryllium-8 does not last long enough to permit any reasonable level of carbon formation by interaction with helium. Finally, the odds of three helium atoms hitting each other simultaneously in just the right way to stick together as carbon is out of sight. It looks like there isn't any way to make carbon. And as noted, without carbon there is no life. Now, it is important to understand that without carbon there is still a universe. It does not even look very different from what we have, on a superficial level. The basic fuel for stars is here, and the stars burn very nicely. The only important difference is that we are not around to enjoy it. But we know that we are, in fact, around. And we know that there is lots of carbon around. So where did it come from?

In 1954, Fred Hoyle of Cambridge proposed a solution. He suggested that there is a resonance between helium-4, beryllium-8 and carbon-12. A resonance describes an effect where one gets a big result from a relatively small effort. Pluck a string in a certain way and you get a big sound for a small pull; do it some other way and it goes flat. Taking into account the mass-energy of each nucleus, and the calculated kinetic energy of the moving particles based on the temperature in the star, Hoyle predicted a hitherto unsuspected energy level, at 7.82 million electron volts, in the carbon-12 nucleus that would cause a resonance for the combined energies of the three elements. This resonance causes three helium-4 particles to stay together just a bit longer than usual, and that is long enough for these to rearrange themselves into the compact and stable configuration of carbon-12. The prediction was tested in the laboratory and found to be correct. The question to ask is, what is it in the basic laws of the universe that requires this resonance, involving three elements and the conditions inside a star, to be there? Why not have a universe without carbon? We do not know the answer. But wait, as they say in the advertisement, there is more.

Once the beryllium bottleneck is solved, there is nothing to stop the process from continuing. Carbon-12 is stable, and helium-4 is stable. There is plenty of time for these to join together and make oxygen-16. We need oxygen. We would not be around without oxygen. But we do not need all the carbon to turn into oxygen. And this would have happened, except for another resonance. This time the resonance is just 1% off in the wrong (or should we say, right) direction. Carbon-12 and helium-4 will not efficiently stick together into oxygen-16 under conditions found in stars. We do have lots of oxygen, but most of it was manufactured for us by primitive bacteria on the early Earth. Having it all cooked in stars would have prevented the formation of carbon-based life. So now we have two resonances without which life could not exist. Why is it this way?

Why does matter exist?

The relationship $E=mc^2$ works very nicely to transform energy into matter and vice versa. The process has been verified in the laboratory in high-energy physics experiments. But you need enough energy to create two particles and not just one. This is because matter created this way comes in two-particle pairs. An ordinary matter particle and its antimatter counterpart. If you have enough energy to get a negative electrical charge electron, you also get a positive electrical charge antielectron, called a positron. A proton comes with an antiproton, and so forth. Therein lies a problem. Matter and antimatter can not coexist. They annihilate each other into energy whenever they get together. Therefore, once the universe cooled below the point when new matter could form out of photon radiation energy, all of the remaining matter particles should have been annihilated by conversion back into energy along with an equal number of antimatter particles.

There are a number of canceling equalities in the universe. For example, the total electrical charge of the universe is believed to be zero. This means that the number of positively charged protons should equal the number of negatively charged electrons. Likewise, given the process of matter creation from energy in the form of particle pairs, there should have been an equal number of particles and antiparticles. At first it was suggested that not all the matter and antimatter annihilated each other, and there is antimatter to be found somewhere. But scientists are now convinced that this is not correct, and that there really is an excess of matter over antimatter. All would have been annihilated, had the two types of matter been equal, and we would not be here. It has been calculated that the excess of ordinary matter particles over antimatter particles is one particle in one billion. All the particles and antiparticles annihilated each other, except that an excess of one out of a billion ordinary matter particles remained. And that is what this universe is made of. Why this exception to the usual parity rules?

There is a clue to the excess of matter over antimatter in the recently discovered "CP Violation," showing an asymmetry in the matter-antimatter relationship in the decay of a particle called the K^0 meson. This leads to a mechanism that provides more matter than antimatter. The reader will find further details in the section "why matter and not anti matter?" in *Just Six Numbers* by Martin Rees. But whether we understand the mechanism or not is not germane to this discussion. The question is why, and not how. Why is there a mechanism in the "nature" of the universe that provides for our existence when the usual rules should not permit this?

All right, so the matter-antimatter equality principle is violated. We end with an excess of matter that makes up our universe. But the conservation of charge is apparently not violated. The universe has just as many negatively charged electrons as it has positively charged protons. The net total charge is zero. Not only zero in the aggregate, but also zero on a local basis such as the earth and sun. This is necessary for us to be around to talk about it. Electromagnetism would not work the way it is described by Maxwell's theory if charge were not conserved. And the electrical forces between the earth and sun would affect the gravitationally controlled orbits if charge were unbalanced by just one part in 10^{36}. Hence, scientists are confident that charge is conserved. But why are there exactly the same number of electrons and protons left over from the initial hot-universe annihilation process?

Why is Earth friendly to life?

There is more that one can say here. Some is fairly specialized, like the resonance coincidences, and some is relatively commonplace, that most people know about. The Earth is at the right distance from the sun, not too hot and not too cold. But there should not be a planet at this location. The planets, excluding Earth and including the asteroid belt, are spaced at a sequence that roughly doubles with distance from the sun. There are many "natural" processes that behave this way, in accordance with what is known as Zipf's law. The mean distance from the sun in millions of miles is 36 for Mercury, 67 for Venus and

141.5 for Mars, for example. So what is the Earth doing at 93 million miles? It does not fit into the sequence of planetary orbit distance ratios. But the Earth is here. And its position is just right for the maintenance of life. The book *Rare Earth* reports on page 19 that the commonly accepted limits on the CHZ (continuously habitable zone) distance from the sun may vary by no more than –5% or +15%. Move the Earth 5% closer to the sun or 15% further from the sun, and we would not exist. Furthermore, the Earth's orbit is highly circular. A more eccentric orbit would not be good for us.

We have a large Jupiter at just the right position and gravitational attraction to catch incoming debris from hitting the Earth. Hence extinction-event impacts on Earth are spaced many millions of years apart. Lessen the interval, and we would not be here. We have a large moon with large tides. Tides are considered by some to be important to life. But the moon is not so big as to turn the Earth-moon system into a double planet orbiting each other. We need and have liquid water. But the water would not be liquid if water behaved like most liquids and solids do. Usually the material shrinks and gets more dense as the temperature drops. Such behavior would cause ice to be more dense than an equal volume of water, and the ice would not float. Water would freeze from the bottom up and not from the top down. Bodies of water would freeze and not thaw out, and there would not be any life.

These examples and calculations are real, and more examples are possible. These are all part of what science tells us. Scientists agree with the information. But that does not mean that scientists necessarily draw the same inferences as someone attuned to the Bible. How, then, does science explain these staggering odds and coincidences?

Science responds

The easiest way to get rid of a problem is to ignore it. That is what many scientists are doing. They pretend, or believe, that these are not scientific but philosophical questions. Hence, they do not need to get involved. Others tackle the problem head-on, as in the book, *Rare Earth.* The authors clearly show how rare the earth is with statements such as, "Earth is a rare place indeed" (page33), "an enigma" (page 45), "Earth indeed may be extraordinarily rare" (page 275). The fact that it is an enigma is acknowledged, but the authors go no further. Others yet, such as Stephen Hawking as represented in his *A Brief History of Time*, readily agree that the idea of outside direction is inescapable if the rules by which the universe operates remain as we understand them today. But the rules do not need to remain as they are today because cosmology science is still a work in progress. These people believe that a better scientific understanding in the future will eliminate the need for outside intervention in the affairs of the universe.

We finally come to those scientists who believe that there is a scientific resolution of these problems already, by means of what is known as the *anthropic principle.* A principle is not itself a scientific law or theory, but is a way of thinking and analysis that points towards the formation of new theories. Hoyle's idea about a resonance to resolve the carbon formation dilemma is sometimes used as an example of anthropic thinking at work. Basically we ask what it would take to make the universe work the way it has to work, given that we exist. This may call sometimes for a rather complicated explanation. The result is sometimes at variance with another well known-principle, known as Occam's razor, which requires that we choose the simplest possible explanation. Hence, not all scientists support anthropic thinking. The *anthropic principle* has its supporters and detractors. Whatever the case, numerous ideas about the universe have been proposed on the basis of anthropic thinking. This includes ideas intended to explain the universe as a "natural" phenomenon and not in need of outside intervention.

There are several versions of *anthropic principle* in use. Two are generally used with scientific precision, and are well-explained in various publications. These are the so called "weak version" and the "strong version." A third version seems to be an outgrowth of the general explanation of what this principle is about. This explanation is usually along the lines that the *anthropic principle* states that we observe the universe that we observe, because if it were different we would not be around to talk about it. One of the useful consequences is to turn things around and ask how the universe must work, in order for us to be here to observe it. This is the idea presented in the previous paragraph. But some people appear to take it as an explanation for why the universe is the way it is. The mode of argument goes somewhat like this. We know with absolute certainty that a universe like the one we have is not impossible. Hence, there is no particular reason why the universe should not be the way it is instead of some other way.

Yes, a different kind of universe is also possible. But then we would not be here to worry about it. This is a pure "circling the target after you shoot" argument. Most scientists who use anthropic thinking recognize the above fallacy. They recognize that the odds are not acceptable and they try to find better odds by using this principle.

There is, obviously, more to it, but here is a brief summary of the weak and strong versions of the anthropic principle. The emphasis, in both cases, is on improving the probability at getting the kind of universe we have, one that allows for beings like us. The weak version says that in a universe that is large in space and time, preferably infinite, there will be regions with different conditions. Obviously, we are in a region that fits beings like us. There may be totally different beings in a different region that are asking the same question about their part of the universe. We believe, currently, that the universe is limited in time. And that this time, in fact, is just about enough for intelligent beings to develop. We could not have come into the picture sooner because it takes a certain amount of time for the first stars to evolve, for supernovae to spew out carbon, for new stars and planets to come in that include this carbon, for simple life to develop, etc. Hence, we could not be here sooner in a universe that is limited in time.

Our observational universe, that is the event horizon, is also limited in space. This is because it takes time for a light signal to get to us, and that time is limited to the roughly 15 billion years since the big bang beginning. But there could be something else beyond the event horizon with very different physical conditions. This idea improves the odds for what we have, but not by all that much. The strong version of the principle, however, improves the odds enormously.

The strong anthropic principle posits that there are either an almost infinite number of different physical regions in a universe of infinite extent, or else that there are virtually an infinite number of different universes. There is no problem with probabilities either way. Whether a googol, or googolplex or a googolplex googolplexes, it makes no difference when compared to infinity. Every physics-permitted universe exists. Originally we were asking, Why do we live in a house when everybody else lives in a hut? In fact, where does this unique structure we call a house come from? But now we ask, Why do we live in this house instead of some other house? The answer is that any house we lived in, among the many houses in existence, would elicit the same question. The strong principle sounds like science fiction rather than science. But the "many universes" idea is taken seriously by some theoreticians because analysis from quantum mechanics shows that it could be possible.

Martin Rees asks in *Just Six Numbers* how this universe can be explained. He concludes on page 148 that there are only three possibilities, "Coincidence, Providence or Multiverse?" Coincidence is impossible because the odds are basically zero. Providence is not acceptable to him. Hence he devotes a chapter in his book to explain why he favors a multiple universe hypothesis. Currently this idea has moved out of speculative science and into the general population. For example, the cover of the July 20, 1999, issue of *U.S. News* magazine asks, "Is Ours The Only Universe?" Those who accept this idea are pleased that the problem with impossible odds that point towards a Biblical interpretation has been eliminated. Sorry, the Torah was there already. Rabbi Judah ben Barzilai, of 12th century Barcelona, comments on the many "worlds" discussed in the Talmud Tractate *Avoda Zorah*. He explains that "world" does not mean a planet but a separate universe. A "many universes" interpretation does make matters easier for the non-believing scientist, but it does not make it more difficult for the science-inclined believer.

Where does the above lead in a comparison of Torah to science? "Coincidence, Providence or Multiverse?" Is a Creator any more speculative than a Multiverse? We see that in most instances, Torah easily explains results that science cannot touch. And if we do wish to accept science fiction-like scientific ideas, such as the "many universes" hypothesis, we find that Torah was already there too. No matter how you slice it, Torah is not harmed by science.

Chapter 4

Science or Philosophy?

Evolution

We now address the most contentious, and according to many people, the most difficult, issue, namely evolution. But the idea that evolution is the most difficult issue, and very dangerous to belief in Torah, is simply not correct. Even classical, gradual, evolution does not threaten basic belief in Torah, though the proponents of evolution frequently portray it as though it does. There are still those who cling to the idea of a slow and gradual evolution among species, but this idea is under strong attack by the more adventurous evolution scientists. There is just too much scientific evidence against it. Rather, a hypothesis called "punctuated equilibrium," involving virtually sudden changes, when measured in terms of geological time, seems to be the direction for future theories. Thus, authors Peter Ward and Donald Brownlee, reporting on the latest findings on page 140 of their book, *Rare Earth*, published in 2000, note that, "It is clear that the evolution of animals occurred not as a gradual process but as a series of long periods of little change, punctuated by great advances." Here the Torah is very secure, since the sudden appearance of new creatures is the plain meaning of what the Torah tells us.

A fast evolutionary transformation

Life, even of the simplest forms, is characterized by complexity and organization. It seems inconceivable that such a phenomenon could arise on its own without some sort of direction, without a designer. Hence the so called "argument from design." This argument was made by Rabbi Akiva nearly two millennia ago, and most are familiar with the more modern version presented by William Paley (1743-1805) in his *Natural Theology*. His analogy of the watchmaker is probably the most quoted example of this argument. Paley asks whether, having found a perfectly assembled and functioning watch in the field, one could possibly do other than infer the existence of a watchmaker. But Charles Darwin's *The Origin of the Species* does challenge that assertion. There is no question but that mutation and natural selection, as understood in their broadest forms, are a fact of nature. Yet the fact that a general process exists does not prove that it can produce a particular result. It can be proven, for instance, that people have at their disposal methods of transportation which are quicker than walking. The existence of certain fast-moving vehicles is a provable fact. But that does not mean that we can extrapolate this to the particular claim of an instantaneous transport of one thousand miles. Not at least, without invoking miraculous intervention.

There have been numerous probability-based analyses of what is, and is not, feasible for Darwinian evolution to achieve within the available time, based on the age of the universe. Most calculations show that evolution does not work this way. But please remember that we are dealing with probabilities, and the result depends on the initial assumptions. A powerful argument in favor of gradual evolution, using somewhat different and seemingly reasonable assumptions, was launched in 1987 when Richard Dawkins published his book *The Blind Watchmaker*. The connection between the title of this book and the Paley watchmaker argument is obvious. Numerous pro-evolution publications quote from this text as from a bible. Indeed, this work seems to have become the current bible of evolution. But rather than a threat, this author sees it as supporting the Torah.

The first item that caught my attention is the definition of miracle used in this book. Thus we are told, on page 139, that that which we call miracles are just "improbable natural events." This is in line with the Torah view, as quoted from Rabbi Dessler, that the miraculous is hidden from us as rare natural events. Dawkins' description is not contrary to Torah. Next consider the specifics. Dawkins points out, correctly, that an artifact, like a watch, is not the same as a biological organism. A watch does not evolve. It must be completely and correctly assembled, to function mechanically. Hence the probability of blind chance assembling the large number of intricate specialized parts, without human agency, is near zero. Biological organisms are different. We have both simple and complex creatures, and many levels of complexity in between. Only tiny changes need be made at any one step, to obtain the final result. The probability numbers are not just different, they are startlingly different. Especially if you choose the "right" assumptions for the gradualist model.

Dawkins uses the analogy of the proverbial monkey typing the phrase, "Methinks it is like a weasel." The probability of that happening in one random selection of letters, what Dawkins calls "single-step selection," is one out of 10^{40}. But suppose instead of trying at random again, the next try involves a slightly improved modification of the first result. A computer simulation shows that we get the desired phrase in just over forty tries using this "cumulative selection" process. This is a most efficient process, and we get the desired result very quickly. But this also involves a "designer" who chose the desired phrase in advance. Torah has no problem with this idea. But Richard Dawkins does, and he quickly points out that "cumulative natural selection is blind to the future." In other words, the content of the phrase is not known in advance. The gradualist process is still highly efficient, provided the desired small changes are locked in place as these occur, ready to accept the next favorable change which comes about purely by blind chance.

Let us consider the implications of this model. Having the desired changes locked in place, while undesired changes can be destroyed on the next try, is crucial. Remember that the improvement at each step is tiny. The likelihood that such a change will propagate throughout a population which is then ready to accept the next infinitesimal change, is essentially zero. Unless, that is, each of the successive desired changes is permanently locked in. This process is somewhat similar to a mathematical computation process that I am familiar with. My technical specialty is in a field called spectrum analysis. One of the procedures in this field is to compute a certain mathematical function known as the Discrete Fourier Transform (DFT). This normally can take many minutes to compute, as it requires N^2 operations for an N point record length. But we no longer do it this way. A 1965 technical paper showed how to break the N point array into smaller sub-groups to multiply, which are then combined by summation. This process, known as the Fast Fourier Transform (FFT), is thousands of times faster than the previous procedure.

This seems a good analogy to what Dawkins has done. I will call it a Fast Evolutionary Transformation (FET). An essential feature of the FFT is that there is an intelligent designer driving the process even though the final result of the computation is not known in advance. The same applies to the FET. How else does each small evolutionary mutation know to be locked in place, in preference to the thousands and millions of changes to be discarded, on its way to becoming a tiger or a butterfly or human being? Dawkins' watchmaker may be blind, but he must be intelligent and conscious. What greater support for the suggestion of intelligent design? Alas, the FET does not provide any proof for a "designer," because pretty much all available evidence shows that, even if there is some validity to evolution in some sense, it does not behave in accordance with the model used by Dawkins. Even gradualist evolution requires many more terms to compute, and the fossil record points to abrupt, punctuated, evolution if we are to fit a theory to the available evidence. Where then does evolution fit in with respect to Torah?

Evolution in Torah

It is a basic premise of these essays that all science is contained within Torah, and there is no conflict between basic Torah and basic science. This assertion holds also for evolution. But we all "know" that this is not so. We all "know" that evolution and the Bible cannot coexist. What is going on here?

Ask a group of people, whether Torah-observant or not, where the greatest conflict with science lies, and the majority will pick evolution. Why is this such a contentious issue? And what is the issue, or conflict in the first place? A careful analysis will show that there is no particularly more serious conflict between Torah and the science of evolution than with any other scientific discipline. The primary conflict, rather, is with conjectures and philosophies based on evolutionary science, but which themselves are not scientific. The matter is further exacerbated when some of these philosophies and conjectures are presented as scientific fact and defended with tenacious zeal. Certainly, those who believe in Torah are permitted, indeed required, to defend the honor of Torah against anti-Torah philosophical ideas. But this has nothing to do with objective science. The theory of evolution, in whatever variation one may consider it, is far from a proven fact. But the scientific aspects of evolution would not threaten the basic validity of the Torah, even if it were shown to be correct. What, after all, does the Torah teach us about the establishment of the universe, living beings in general and mankind in particular?

The universe, including time, space and all matter and energy was brought into being in one unique act of creation (Hebrew, *barah*). Subsequently already existing, previously created, material was made (Hebrew, *oseh*) into various physical structures, such as stars and planets. Eventually, living beings, including a human-shaped being, were formed (Hebrew, *yatzar*) from the existing materials. All of these had the same status as living beings, or animals (Hebrew, *nefesh chayah*). But the human-shaped being was soon made unique as a special, a human, soul (Hebrew, *nishmas chaim*), was breathed into it.

How is evolution, of any variety, supposed to falsify this account? Living bodies of human form are shaped and molded out of previously existing materials. The account is terse and the process mysterious. Suppose the shaping and molding was done through a process of mutation. So what? The Torah teaches us over and over again that the world operates in a manner that is intended to look "natural" and understandable through human reason. This is so we can have free will to believe whatever we wish, even in a form of evolution that posits the possibility of a full change in only a few tens of generations. Rabbi Luzzatto summarizes this in his early 18th century work *The Ways of G-d.* He explains that the physical universe is so arranged that mankind "exist[s] in the present world, bound and limited by its natural laws .... Man was created with a Good Urge (*Yetzer Tov*), and an Evil Urge (*Yetzer HaRa*) and free will." Andrew Goldfinger explains, in his *Thinking About Creation*, how a slightly different spelling of *yatzar*, in the shaping of human beings, leads to our understanding of the two inclinations of people, to good and to evil, as quoted above. The reader will also find a detailed discussion on the implications of this variation in spelling on page 61a of Talmud Tractate *Berachos*.

The science of evolution, of whatever variety, has nothing to do with good or evil inclinations. It has nothing to say about free will or human consciousness. The science of evolution does not teach us how to exercise our free will, what to choose and what to reject. It has nothing to say about a human soul that distinguishes a human being from a human-shaped animal. It deals only with the detailed mechanics of the structuring and shaping of physical matter into a specific physical structure, be it fish, or horse, or human. The Torah too tells us that these were shaped out of existing matter. Certainly, one should disagree when one believes that conjecture is forcefully misrepresented as scientific fact. But, as stated previously, there is no inherent conflict even with the most absurd statements about evolutionary facts. The issue between Torah and evolution is with evolution as a philosophy rather than evolution as science.

Evolution – science or philosophy?

There are three broad stages of evolution: evolution of inanimate matter into life, evolution of the fundamental structure of life, the gene, and evolution of new species. In each case we see that the word evolution is intended as a short-hand name for a description or explanation for what would ordinarily be termed "creation." It suggests also a process of change over time involving a mechanism of mutation and natural selection. Let's say that a gene, subjected to bombardment by fast atomic particles from radiation, changes to a different gene after being hit by these particles. The new gene did not exist before, but it now exists. We can say that a new gene was created. Let's say a "soup" of inanimate matter is subjected to various physical and chemical processes, and living matter emerges. We now have living matter where before there was nothing living. Has not life been created? Nothing resembling a giraffe existed in the world previously, but now the world is full of long-necked giraffes. The giraffe, with all the special mechanisms needed to maintain the blood flow along an exceptionally long neck, has been created.

The concept of evolution is not an alternative to, nor is it in competition, with creation. Evolution is intended to provide an explanation for creation by way of a natural-means process. Torah does not object to natural-seeming processes. Of course, we are not obliged to accept this theory. But it is important to understand that evolution theory, in its strictly scientific mode, cannot falsify Torah.

The real problem comes in when non-scientific evolution theory is treated as if it were the science of new creation via evolution. Evolution is supposed to introduce change through the mechanism of random mutation, with the most suited surviving, while the less suited to the environment do not survive. This is known as the survival of the fittest doctrine. Possibly the most publicized example of this doctrine involves the *Briston betularia* moths in Manchester, England. Over a period, of about a century, industrial soot darkened the trees in this area. During this time period light-colored moths virtually disappeared and were replaced by dark-colored ones. There is no disagreement as to how this came about. When the trees were light-colored, predators could easily catch the dark-colored moths but not the lighter ones. The advantage for survival shifted when the trees became dark-colored. This is an excellent example of adaptation, a process that can be easily verified in laboratory experiments. But this is certainly not an example of evolution whereby a more complex organism evolves from a simpler one. Given the lack of genetic testing in those days, we do not even know whether a new gene, favoring a dark color, emerged or whether previously existing dark-colored moths were now able to breed unhindered while light colored-ones were eaten. Either way, no new organism is created here. Previously the dark-colored moths were eaten, and later the light-colored ones were eaten. A moth is still a moth. Yet the fact that survival of the fittest works in this case, is used as "proof" that it applies to evolution on a broad scale, involving mutation and the bringing forth of something that did not exist before.

No one questions the occurrence of genetic mutations, which is a fact and not a theory. And we do not need to extrapolate from moths for this conclusion. The effort to decipher the human genome has made startling progress. We now have all of the roughly 100,000 genes $^{\Delta}$ on our 23 chromosome pairs identified, located, and recorded in a recipe book. Written in a four-letter alphabet consisting of A (adenine), T (thymine), G (guanine) and C (cytosine), the total comes to about 3.5 billion letters. It is quite a substantial-sized recipe book that would take more than a dozen years to read on a continuous basis. Most of the code apparently does not do anything. Or maybe it does something that we do not yet recognize or understand. Consequently, the vast majority of the thousand or more letter combination transpositions and errors that each person has seem to be benign, without any effect. But some errors are known to either predispose one or directly cause the nearly 4000 inherited diseases that we know about. For example, a simple substitution of a T for what should be an A on chromosome 11 causes sickle-cell anemia. Change is either neutral or deleterious, which is the usual situation for any complex organism or machinery. Where are the helpful mutations? We do not know. As to turning a human being into something else entirely, we are beyond the "we do not know" stage. All we know is that the odds of getting there by blind chance seem to be beyond the googol range.

Looking at the evidence, as opposed to conjecture, we find that there is some form of evolution going on involving survival of the fittest mutations. Whatever the process, the evidence works against the classical Darwinian gradualist model. At best we have to deal with abrupt changes, and at this time no one understands how the process works. Evolution of species is supposed to be a fact. But the fossil record does not support the contention. The response is that the record is not complete, and we will find the stuff if we dig long enough. We cannot disprove this idea by absence of information that may become available later. But we can insist that the idea is just that, an idea, till the fossil record supports it.

Of greater significance is fossil record information that is available. New life appears to emerge fully formed from out of nowhere in virtually zero transition time. How did evolution accomplish that? We do not know. What are the mechanisms that drive evolution? We do not know. Ward and Brownlee, authors of the much talked about book *Rare Earth* take evolution as a given. Nowhere in the book is there the slightest hint that some other process, such as a Creator, might be involved. What words or phrases do they use in describing how evolution works? The book is filled with such words and phrases as "fortuitous," "may have been," "possibility," "unique event," "extraordinary conditions," "imply," "may be," "hypotheses," "possibly," "probably," "belief," "would have," and "may have." They conclude on page 37 that "limited by natural laws and materials, it is unlikely" that conditions of planet Earth and life can be duplicated. So how did life come about? How is it that we are here? Science has a theory. It is called evolution. But when we get down to details, the simple answer is that at this time we do not know. This is an answer that science is not happy to give.

Originally evolution theory posited that life developed slowly over a billion or more years by reaction of existing chemicals with the environment. Experiments nearly fifty years ago appeared to confirm this as amino acid, precursors to life, were created in the laboratory. But we have not heard much about this in the last twenty odd years. It was a good idea, but it does not work. One important contrary point is that we know now that life formed virtually immediately with the appearance of water on the planet at 3.8 billion odd years ago. Nobody understands how this came about.

The middle stage in evolution involves new gene formation. Certainly there are mutations. But to get to anything useful involves probabilities in the googol range. Unless, of course, one makes the kind of assumptions that Dawkins made, as discussed previously. Whatever "natural" seeming process will eventually be found to explain life, including humans, it will not be the neo-Darwinian idea of gradual evolution. The point is that the scientific evidence inherent in the fossil record, chemical processes and mathematical computation is in accord with the results described in the Torah. The choice of vehicle to carry out the result is not ours to debate. As noted at the outset, evolution science is not at variance with Torah. Only evolution philosophy is.

There are different names for this sort of non-scientific thinking that can grow around a body of scientific thought. Some people call it scientism. I have chosen to call it a philosophy. But given the zeal with which adherents defend their position, it might as well be called an alternate form of religion. Whatever the name, it is important to remember that it is not science. Evolution scientists work under a handicap compared to those in other scientific fields because much of what they propose is not testable by experiment. Hence, the field is full of speculation and conjecture, almost by necessity.

But all is not lost, because we do have the fossil record. Evolution is similar to cosmology in this sense. We cannot go back in time to observe the formation of the sun or perform an experiment in planetary formation. Neither can we go back in time to examine a living dinosaur. But we can examine the record of the past in both cases. These disciplines are also similar in that there are associated scientific endeavors which can be tested by experiment. This might be high energy physics for cosmology, and biochemistry for evolution. Also, both areas started out in a highly speculative and philosophical mode with a gradual transition to more and more hard science. Finally, for whatever reason, both disciplines find themselves at apparent odds with the position of the Bible. Cosmology transitioned from a highly speculative state about forty years ago to a firm science methodology footing today. But evolution is not there yet. And this is a prime cause of much misunderstanding. Let us examine a bit of the history of cosmology and compare that to the current situation in evolution.

An example from cosmology

Lacking experimental or observational evidence, it is possible to theorize in any desired direction. The only constraint is that it has to be not impossible. Cosmologists were busy with many ideas fifty years ago, but there was not much firm progress in narrowing down the possibilities. Virtually all involved in the science would insist on a permanent universe if asked. Why? Was the evidence for this model so compelling? Actually, no. The evidence for a definite beginning involving the dark sky at night, known as Olbers' Paradox and an expanding universe, pointed to the opposite. A theory by Gamov for a definite beginning, a theory in its infancy, but well-structured nonetheless, was there to consider. Still most cosmology scientists refused. It was not till 1965, with the discovery of the cosmic background radiation (CBR) residue from the "big bang" start of the universe, that the cosmology establishment was forced to agree that the universe had a definite beginning. Why? Says Nobel-winning physicist Steven Weinberg in his book *The First Three Minutes*, because a steady-state universe avoided a problem with Genesis.

Who said that scientists dispassionately respond only to facts and do not act on emotions? Acknowledging anything that even remotely smacks of a validation of the Bible was apparently not easy for most mainstream cosmologists. And the emotional dimension appears to be even stronger in the case of evolution.

Why do evolution scientists feel so strongly about their discipline? For that matter, why is this such a hot-button topic for all of us? There was a survey of experts done in late 1999, and a list was compiled of the 100 most influential people since the year 1000. Darwin was number four on the list, ahead of Newton, Einstein, Mozart and Shakespeare, to name just a few. Newton was enormously prolific. He made important contributions to just about any scientific discipline one can name. And his work was so correct that we do not talk about Newtonian theories but Newtonian laws. If a leap of intuition into the unknown is to be the measure, then surely Relativity cannot be easily dismissed. And who among us would suggest even the tiniest changes to Mozart's music or Shakespeare's words? Not even the staunchest neo-Darwinian would today insist that Darwin's original conjecture can stand unadjusted. So why is Darwin ahead of these people? The experts who made the choice explain that Darwin was dealing with the most important topic we can talk about, ourselves.

Ourselves involves science, but ourselves also goes beyond science. Who we are, what we are, why we are, who and what and how we want to be. All of this, and more, is part of ourselves. The Torah says that we share the condition of having a body with all animals, but it is the condition of having a human soul which makes us different. Can evolution science find the soul? Certainly not yet. Hence, we are told that human beings act in certain ways, or should act in certain ways, because we are descended from, or at least cousins, to the apes. This is not evolutionary science or any kind of science. This is philosophy or religion or whatever, but not science. This is where the irreconcilable differences lie, and this is why science will not easily change to accommodate the plain observational facts. Because a change in the science would involve a change in philosophy.

There is a legitimate place for the examination of differing belief systems. But this essay is not it. Here, the focus is only on Torah and science. And those who believe in Torah are firmly convinced that eventually we will have full agreement, when evolution facts catch up with the evolution philosophy currently pretending to be a science.

Δ The latest information, as we go to press, indicates that humans have less than half as many genes as was once thought. The number is not more than 40,000 and not 100,000. No one, as yet, understands the implications of this surprising discovery.

Chapter 5

Why Is the Sky Dark at Night?

The Universe

Why is the sky dark at night? This appears to be a childish question with an obvious answer. The part of the Earth we are on at night is facing away from the sun. The light of the sun cannot penetrate the Earth. The only light left is reflected from the moon and planets and the feeble bit of light reaching us from distant stars. That is why the sky is dark at night. True, but serious scientists like Keppler and Olbers would not have asked this question if it were that simple. This question came to be known as Olbers' paradox, after Heinrich Olbers popularized it in an article in 1823. By the way, Olbers' suggested solution to the paradox, involving intervening interstellar gas, does not work. Nothing does the job fully, except the assumption that the stars had turned on at a finite time in the past. The dark sky at night clearly points to a universe of finite time duration. Why?

Consider a universe of infinite extent in both space and time. Assuming that the Earth is not special in some way, then the universe would appear to be isotropic and homogeneous. This says that the universe would look essentially the same in all directions when observed from one location, and it would also look the same from all locations. Hence, any large enough volume of space has, on the average, the same number of stars as any other equal volume of space. On the average, any star is as bright as any other star. We now calculate as follows. The light intensity from any source is reduced as the inverse square of the distance from that source. This well-known law is very easy to demonstrate. All of the light emitted by the source will be intercepted by the surface of a sphere drawn around the source. The surface area of the sphere increases as the square of the radius of the sphere. Hence, the same amount of light is spread out over an area that increases as the square of the distance, and the intensity per unit area decreases as the square of the distance. This is why far away objects appear to be very dim, even when these are intrinsically very bright. But we are not subjected to the light from just one star. We are subjected to the light from all the stars. The volume of the sphere increases as the cube of the distance. And the number of stars also increases as the cube of the distance. The intensity from any one star decreases as the square of distance. The total stars we are subjected to increases as the cube of the distance. And the total intensity of all the stars, taken as an aggregate, increases directly with the distance. An infinite universe in space and time has all the distance you like, and all the time needed to let the light from even the farthest stars get to us. The sky in such a situation should be blazing as brightly as on the surface of the sun. So why is the sky dark at night?

Olbers suggested intervening dust. But that does not work. The dust would eventually get as hot and bright as the sun, and the process would start over again. Remember that an ageless universe has all the time that is needed. One can suggest that the universe is not homogeneous or isotropic. That the Earth has some special status, and the universe has unique properties in our neighborhood. This idea is usually discarded on the basis of observational evidence that points to the contrary, and on the philosophical grounds that a special status for Earth points to deliberate design. It is interesting to note, though, that the strong anthropic principle proposes a more sophisticated version of this idea, as discussed in Chapter 3. An expanding universe accounts for some of the result because the recession of distant stars reduces their apparent light intensity. But calculations show that this is not enough to give us a dark sky. There is only one solution that takes care of everything – the universe has not been around long enough to cause this effect.

How old would the universe have to be to get an Olbers-type hot sky? Imamura of the University of Oregon made this calculation, which appeared in April of 1999 on the Internet (web site URL: www.zabu.uoregon.edu/~imamura/209/apr7/olbers.html/). He estimates that there is one star per cubic parsec in the universe. A parsec is an astronomical unit of distance equal to 3.26 light years. Taking the radius of the observable universe at 15 billion light years [Δ], we find the volume of such a sphere as 4×10^{29} cubic parsecs. Hence, that is the number of stars used in the calculation. The bright-sky age calculation shows that we get a hot sky condition at a universe age of 6.6×10^{15} years. This is four hundred and forty thousand times older than the current age. No one knows precisely how many stars there really are in the universe. But the lowest estimate of the number of stars in the universe is on the order of 10^{20}. Hence the Imamura calculation appears to be generous. And that means that we may have to wait even longer than the 440,000 times the current age calculation to get to a hot and bright sky. The universe seems very old when stated in absolute terms of billions of years. But it is still very young when judged on a comparative scale with all cosmological events.

For whatever reason, scientists did not use the dark sky to embrace a finite time start for the universe. The idea, after all, was purely theoretical without any observational evidence. Enter the expanding universe. In 1913 astronomer Vesto Slipher noted that some nebulae were receding from us. He continued his work well into the 1920s, but the idea of an expanding universe did not catch on. In 1916 Albert Einstein published his monumental Theory of General Relativity. It was soon discovered that the mathematical solution involving a static universe was not stable. First, Danish mathematician Willem de Sitter suggested a solution that appears static but includes a recessionary red shift term. Later Soviet mathematician Alexander Friedmann proposed a solution involving an expanding universe.

But Einstein, along with other mainstream investigators, wanted a static universe. Einstein stuck with his solution of a static universe that is stable, by the addition of a term called a cosmological constant. Later, when an expanding universe was no longer in doubt, Einstein averred that the cosmological constant idea was a big blunder. Finally, in 1929, the investigations of Edwin Hubble made it clear to all that the universe was indeed expanding. Obviously, an expansion means that the universe gets bigger as time passes. Just as obviously, one can conclude that the universe was smaller in the past. Indeed, if we follow this reasoning far enough into the past the universe gets to be of zero size, implying a beginning. But no, mainstream science refused to come to this conclusion.

A universe of finite age continued to be resisted even though this introduces some difficulties. The second law of thermodynamics requires that the degree of disorder, identified as a quantity called entropy, must increase within a closed system. Highly organized entities, such as living beings or planets or galaxies, must deteriorate in orderliness as time goes on. We can increase order on a local basis, but the effort to do it always causes more disorder overall in the total closed system. The universe is a closed system. And it is currently highly ordered. How can that be, if the universe has been around forever? The universe should have totally deteriorated into stray wisps of particles and energy if it were very old. But we know that this is not the case. The obvious conclusion is that the universe is of finite age. That too was not accepted as a sufficient reason to take a beginning, a "genesis," point seriously. Yes, there were scientists, like George Gamov, who championed such an idea. But the conventional wisdom said, no.

Never underestimate the ingenuity of scientists. There is always a way if an idea is attractive enough. And an ageless universe, or should we say the avoidance of a finite age universe, was most attractive. The answer was a steady-state theory built around continuous creation. The fact that this theory was championed by Sir Fred Hoyle, the same one who solved the carbon-beryllium bottleneck discussed in Chapter 3, did not hurt the credibility of this idea. It is a simple, yet brilliant, idea that solves many of the problems with an ageless universe in one go. Continuous creation postulates that new matter is continually created. The amount per unit volume and time frequency is extremely low. Hence, we could not observe this with any technology currently known. Every once in a while a new particle just appears out of nowhere. This provides for an expanding universe that has existed forever, and it answers the questions about the second law of thermodynamics.

No, Sir Fred did not suddenly start writing science fiction masquerading as science. Quantum mechanics provides the underpinnings for such an idea. Indeed, the idea survives in the current proposals for how the big bang universe began. The steady-state theory was tenaciously defended, no matter what. Why? Says Nobel prize winner Steven Weinberg in his book *The First Three Minutes*, because "the steady-state model, ... nicely avoids the problem of Genesis." Weinberg explains that "it was extraordinarily difficult for physicists to take seriously any theory of the early universe." But everything changed in 1964-65 with the discovery of the cosmic background radiation (CBR). And now the heretical big bang idea has become the standard model. We will get to this development in a moment. But first a few words about measuring distance, time, the red shift and the expanding universe.

Distance, the red shift and expanding universe

It starts with a search for distance information. How far is one place from another place on earth? How large is this planet Earth? What is the distance from the Earth to the moon? How far is the Earth from the sun? Once we know the radius value as the Earth orbits the sun, it becomes possible to measure the distance to much more distant objects by parallax triangulation techniques. We form a baseline using the diameter of the Earth's orbit as the Earth changes position at a half-year interval. The apparent shift in position of a nearby star against the background of fixed or distant stars, known as parallax, can then be used to calculate the distance to the nearby star. But even the closest stars are enormously distant from us. The light from our closest neighbor, Alpha Centauri ΔΔ, takes over four years to get to us. The parallax for this star is only 0.761 second of arc. A frequently used astronomical unit of distance, the parsec, is defined on the basis of parallax. A parsec is a distance of 3.2615 light years, or 3.0856×10^{13} kilometers, where the parallax equals one second of angular arc. No wonder the first stellar parallax measurements were not obtained till 1838 when Friedrich Wilhelm Bessel, best known for the mathematics of Bessel functions, measured the parallax of the dim star, 61 Cygni. Today with our immeasurably more precise instruments, we still have less than ten thousand stellar parallax measurements. For a while it appeared that we would be limited to measuring the distance to just our closest neighbors. But then we got into a whole new ball game with the discovery of Cepheid variables.

Cepheid stars are designated as variables because the luminosity varies with time. The important part is that it was found that the luminosity exhibits a close relationship to the time variability. Determine the manner in which the brightness of a Cepheid varies with time and what the color is, and the absolute luminosity is also determined. Now it became possible to determine distances by comparison of computed absolute luminosity and measured apparent luminosity. This not only became possible in our own galaxy, but also in relatively nearby galaxies, as long as an individual Cepheid variable star could be isolated and observed. The process is conceptually quite simple. A star can appear dim, with a low apparent luminosity, either because it does not emit much light representing low absolute luminosity, or because it is bright but distant. The inverse square law tells us how luminosity varies with distance. Hence, the ratio of absolute to apparent luminosity readily yields the distance.

Over time the results obtained in this way had to be readjusted as the variability-luminosity values involving such stars were recalibrated. Today we hold that our closest galactic neighbor, the Andromeda-Nebula, is just over two million light years distant, as opposed to the very first calculations at 900,000 light years. By comparison, our own, Milky Way, galaxy is 80,000 light years in diameter and 6,000 light years thick. It is also surrounded by a spherical halo of stars with a diameter of about 100,000 light years. Hence, Andromeda is twenty times more distant from us than the furthest separation inside our own galaxy.

Millions of light years sounds like a very large distance. Imagine a distance where it takes light millions of years to get to us. But it is not all that much compared to the whole universe. Andromeda is, in fact, part of our local group. Its movement has nothing to do with the movement that is due to the expansion of the universe. The movement is just part of a so called "local motion." And we find that Andromeda is moving towards us and not away from us. We need to go much further to get any appreciable distance across the universe. But here checking on Cepheids, or any other individual stars, becomes impossible because individual stars cannot be resolved or observed at such distances. We cannot observe or make measurements on individual stars, but we can make measurements on individual galaxies. The procedure is the same, but now the luminosity of a whole galaxy is used in the calculations. As previously, overlap measurements using two procedures, Cepheid and galaxy luminosity in this instance, can be used to calibrate the longer distance procedure against the shorter distance procedure. There matters stood till Hubble announced the result that is now known as Hubble's law.

Others had reported what appeared to be a systematic recession from us for distant objects. But Hubble went one step further. He claimed that the rate of recession was linear with distance. Later examination of the raw data shows that Hubble was premature. The results show a progression, but there is much too little information to decide on a linear relationship. Perhaps it was intuition, or even a lucky guess. It makes no difference, because Hubble was right, and his law is now firmly established in the field of Cosmology. Not only that, but a number of important cosmological matters have acquired the Hubble name. There is the Hubble law, the Hubble constant, the Hubble program. And the first space-based telescope is named after Hubble. Why is the insight that leads to Hubble's law so important? To understand this properly we need to consider the matter of spectra in general, and the spectral shift to the red in particular.

The spectrum and Doppler effect

Light can be treated as a wave phenomenon. The color that we see or measure, even if not visible directly to us, depends on the wavelength. The distribution of these wavelengths, or the inverse which is frequency, is called the spectrum. Not only visible light, but all electromagnetic radiation exhibits a spectrum of different wavelengths. Gamma rays and X rays exhibit very short wavelength. Then comes ultraviolet light which we cannot really see directly. Then comes the first truly visible light at a wavelength of 4000 Angstrom units, followed by blue, green, yellow, orange and red with a long wavelength of up to 7500 Angstrom, and where an "A" unit is 10^{-8} cm in length. At longer wavelengths comes infrared, microwaves and then regular radio wave signals. In 1802, William Wollston noticed a distribution of bands, or lines, across the spectrum from sunlight. Later Fraunhofer, Kirchhoff and others refined these results.

Today we know that these spectra consist of both bright lines and missing, so called absorption, spectral lines. The bright lines come from the burning of particular elements at particular temperatures, while the absence of some spectral components comes from the absorption of that light by a cooler outer envelope of material consisting of certain elements. Different elements exhibit different absorption line positions and structure determined by the atomic arrangement of that element. The fact that there is a spectral structure involving multiple lines, is useful in recognizing which element we are dealing with, even when the apparent spectrum position is shifted from where it should be. Different stars, at different temperatures and different element abundance, show different spectrum appearances. Super high temperature blue-color stars, identified as of "O" class, burn at over 25,000 degrees and exhibit spectral lines characteristic of ionized helium and ionized metals. In the center of the range are the white/yellow G class stars that burn at 5000 to 6000 degrees with lines characteristic of both ionized and neutral metals. Our sun is a G type, main sequence star. The coolest M type, red stars, burn at under 3500 degrees. These start to exhibit spectra due to atoms combined into molecules.

The combined spectrum from a whole galaxy is a combination of all possible stars and absorption lines. Not only the particularly prominent lines mentioned above, but many other absorption lines will be exhibited by this spectrum. Helium, for example, was first identified in 1868 as a separate element through spectroscopic analysis of the spectral lines from our sun. More than 60 elements have been identified as present on the sun through such analysis. Having a whole mixture of every possible spectral line is fine for our purpose, because we are not looking to classify specific stars but rather to determine the wavelength position of the measured spectra. This is fairly easy to do with high precision. We look at the fine line structure of a magnified part of the spectrum and identify which absorbing element this represents. Knowing the element, we know precisely what spectrum wavelength this should be. A comparison to the measured wavelength shows by how much the spectrum has shifted. Why care about spectral shifts? The answer is the Doppler effect.

Suppose a source of light moves towards an observer. Successive wave crests and troughs will get to us sooner than they would for a stationary light source, and the apparent wavelength of the light will get smaller. This would represent a shift of the color of the light towards the blue end of the spectrum. A source moving away from the observer would cause the wavelength distance to stretch, which represents a shift towards the red end of the spectrum. Not only the direction of shift, but also the amount of shift can be accurately predicted on the basis of the relative speed of the light source and the fixed velocity of the speed of light. The accuracy of the formula was verified in 1868, 26 years after the prediction of the effect by Christian Doppler. This effect applies not only to light but to any wave phenomenon, including sound. We experience the Doppler effect in sound waves on a daily basis, any time we note the change in engine pitch for approaching and receding automobiles.

Other astronomers had noted a shift towards the red of various distant celestial objects, indicating that these are moving away from us. The critical addition that Hubble made was to claim that the shift was proportional to distance. This has several implications. The most important is that it would appear that everything is moving away from everything else, which implies an expanding or exploding universe. If that is so, then we can use the red shift as a distance indicator for the more distant parts of the universe that we cannot determine in any other way. Hubble's original measurements and results were meager, and his conclusion on shaky ground. But additional measurements by Hubble and others soon verified the conclusion which is now known as Hubble's law.

The rate of expansion of the universe is specified by the Hubble constant, H_0, which is the ratio of the recession velocity to distance. Soon even Einstein, who was most reluctant to accept an expanding universe and added a special term to his famous Relativity equations to avoid it, agreed that indeed the universe was expanding. The obvious conclusion for an expanding universe is that the universe was once very small, and the universe had a beginning. But this was philosophically untenable for the scientific establishment. Instead the expansion was explained on the basis of an ageless universe that is continually renewed by the infusion of newly created matter. Then came the discovery of the currently remaining remnant of the original start of the universe, known as the cosmic background radiation (CBR), and a universe with a beginning in time could no longer be denied.

The cosmic background radiation

This part is particularly interesting for me because my scientific work involves measurements of a similar nature. To be sure, the equivalent temperatures involved are at least an order of magnitude higher, but the idea is the same. Thinking back to 35 years ago, looking at the literature and the available technology, it is clear that even someone of my relatively modest standing in the science community could have performed the measurement before 1965, assuming that money was available to pay for the equipment. Yet the discovery, for which Arno Penzias and Robert Wilson received the Nobel prize, came as a fortuitous accompaniment to a different investigation. A group was just starting out on a deliberate hunt for this information, but the Penzias and Wilson discovery beat them to it. Apparently, those who knew the measurement technology did not realize that there might be something to measure, while those who knew the theory did not realize that the measurement could be made. The measurement relates to an extremely low temperature, but the measurement itself is not of temperature but of random signal noise level. What does a radio signal level have to do with temperature?

Contact between a warmer and colder body results eventually in equilibrium, where the warmer substance cools down, and the originally cooler substance heats up. Heat appears to transfer, or flow, from a hotter to a colder area. For a while it was postulated that heat was a substance, or fluid, that could be transferred between locations. But attempts to isolate, measure and weigh this fluid ended in failure. Heat was eventually explained by the kinetic theory which shows that heat is simply a manifestation of energy. For example, the random movement of gas molecules is experienced as the heat of the gas. The same concept and phenomenon apply also in other circumstances. A random jiggling of electrons can be understood as a form of heat, for instance. The movement of electrons is an electric current, and a current is a signal. But a random current does not contain any information, it is simply noise. Hence, it is common practice at radio frequency and microwave frequency electronics to talk about the equivalent noise temperature, or just the temperature. This is the temperature of a circuit that would cause the circuit to exhibit the observed random noise level as a result of the thermal agitation of electrons. The same concept applies to both conducted and radiated measurements. It is quite common to talk about the noise temperature of the sun, which will vary with sun spot activity. We also talk about the noise temperature of heavenly bodies, such as the planet Jupiter or the noise temperature of the sky. Mostly, this is a convenient shorthand way of designating the amount of a low-level electrical noise disturbance. In the case of interest here, though, the temperature designation has a literal meaning.

Penzias and Wilson, working at Bell Telephone Laboratories at Holmdel, New Jersey, in 1964-65, found an extra bit of unexplained extraneous noise while investigating the transmission of signals at a microwave frequency of 4.08 GHz. Some people would have just ignored this very tiny discrepancy. But to their credit, they did not. They went to all sorts of effort to eliminate or identify the source of this noise. At one point the giant horn antenna used in the experiments was completely dismantled and cleaned of any residue remaining from some pigeons that had roosted there. Eventually they realized the significance of the fact that the noise appeared to be independent of time of year or direction in the sky. After exchange of information with a number of people, it was concluded that this result fit the predicted residue of the big bang beginning of the universe. Penzias and Wilson published a paper showing what they had done and the results they had obtained, while theoreticians Dicke, Peebles, Roll, and Wilkinson, who had just embarked on a project to seek this signal, published a paper explaining the cosmological implications. This signal has since been tested and verified as meeting all criteria as the big bang remnant. It is an exceedingly small signal at only 2.73 degrees Kelvin above absolute zero temperature. The usual shorthand designation is as the 3 K CBR – that is cosmic background radiation – signal. The current value is only 3 degrees, but it started at enormously higher temperatures.

We know that the temperature and pressure was high, beyond description, at the start of the universe. But no one can say with any assurance what happened at the very beginning, because current theory has difficulties at less than the Planck time of 10^{-43} seconds, where quantum mechanical effects preclude a discussion of time in the usual sense. Most investigators don't feel too comfortable with even more time than that. Thus, Weinberg starts a description of the universe at a lengthy one hundredths of a second after the start, in his book *The First Three Minutes*. But that is good enough to get an idea of the incredible conditions of the very early universe. The temperature was 10^{11} degrees at that time. Soon the radiant energy and mixture of radiation and matter in thermal equilibrium cooled off as the universe expanded. This is the same cooling effect that a refrigeration system experiences as a gas expands to occupy a larger volume. The same matter and radiation quanta have more volume to occupy, so the temperature, which is the energy per unit volume, goes down. It is a straightforward linear relationship, hence the temperature drops inversely as the volume of the expanding universe increases.

Today the temperature has dropped to a frigid 3 degrees. But it is all of the stuff as at the beginning. Now all that is left is a very low level of random radiant energy, particle matter having long ago decoupled from the radiation. But that does not matter. Matter and radiation were in thermal equilibrium, and can be treated as equal at the temperatures and energy levels of the early universe. What is a simple, single word, that describes such a condition? Says Weinberg, "the universe was filled with light." But how do we know, one might ask, that the observed 3° K radiation is a remnant from the early universe? For this we need to understand something about the properties of black body radiation.

The fact that the 3° K noise seems to pervade all of space, and come equally from all directions, is a good clue to its origins. Still, having it follow black body behavior would pretty much make the origin certain. Measurements since the first investigation show that this low-temperature noise does indeed follow expected black body behavior, not only in the region of the easier-to-determine longer wavelengths, but also in the short wavelength region dominated by quantum mechanical effects. The energy per unit volume for a perfectly absorbing black body, or a perfectly isolated box, follows a specific energy as a function of wavelength distribution for each temperature. With a mixture of energetic and hot radiation and matter particles in thermal equilibrium, the early universe is like a black body box whose energy distribution should follow black body rules. Matter, of which stars, planets and ourselves are made, eventually uncoupled from the radiation at cooler temperatures and followed its own way. Matter eventually clumped together by gravitational attraction, but the radiation continued to expand with the universe and got colder; yet all the while it followed black body behavior.

The good part about black body behavior is that the intensity will peak at a specific wavelength and fall off steeply on both sides, all determined only by the temperature. Once we have a temperature, it is simply a matter of going through the measurement to determine whether it does or does not follow black body behavior. To be sure, multiple wavelength measurement involving such a low temperature is quite difficult, especially on the short wavelength side. But enough measurements have now been made that there is no doubt about the black body nature of this signal. Not only has this noise signal been determined to follow the black body rules, but the uniformity with space direction orientation is also within the expected parameters of the calculated residue from the big bang beginning. Measurements using the *Cosmic Background Explorer Satellite* (COBE) show that the radiation is remarkably uniform, with just enough tiny deviations to account for the slight differences in temperature caused by the aggregation of matter into galactic clusters. It is not possible to deny the obvious conclusion any longer. The universe had a beginning. Today this idea, which was previously considered a cosmological science heresy, is known as the big bang theory or standard model.

Torah and the big bang

Obviously, the Torah says that there was a beginning, and so does the big bang theory. But the timing of the two is very much different. The Torah deals in days of creation and just a few thousand years of existence, while the standard model deals in millions and billions of years. The gulf might seem too great to bridge, but that is not the case. I devote a separate chapter to this important matter and other matters involving the age and structure of the universe. Here I would just like to show the remarkable concurrence between Torah and science in one small area.

At first there was darkness. This is not the darkness which is an absence of light, but rather an essential property of the universe which was created out of nothing as part of the initial creation. We note in Isaiah 45:7, that darkness was created, as in "Who forms light and creates darkness." Creates, *barah*, is the same word as the second word in the book of Genesis, *Breishis barah*. Later on the first day came light. In prior years, before science caught up with Torah, those who oppose the Bible pointed to this as proof that the Bible makes no sense. How can there be light on the first day when the sun did not come into being till the fourth day? Our commentators also know how to count the days. They point out that this light did not come from the sun, though it is ultimately connected to the sun. Not only that, but it is noted that the light was impossibly bright. Nothing could survive in it. What nonsense, a scientist of even a hundred years ago would say. But not anymore.

Now we have a winner of the highest award in science, the Nobel Prize, stating in a book about science that "the universe was filled with light" (Weinberg in *The First*

Three Minutes, page 6). And the light was indeed bright, at a temperature of millions and millions of degrees. Nothing resembling our physical universe or ordinary matter could survive. The light had to be attenuated, and the temperature had to cool down. And that is what Torah tells us – but not a few decades ago, but millennia ago. No wonder this type of teaching had to be kept in strictest confidence, to be revealed only to those of the highest integrity. What ordinary person could understand or believe such strange stuff? Hence, mainstream commentaries, such as the primary references by Rashi, without which Torah cannot be studied, merely point out the unusual nature of this light without going into specific detail. We get a hint about the process from Rabbi Munk on page 100 of his *The Wisdom in the Hebrew Alphabet*. Rabbi Munk explains that a slight change in spelling of the Hebrew word meaning "light," "is expounded to imply that the original light was withdrawn ... The missing light was replaced by the radiation of luminaries. In the first verse, the root of *meorot* (luminaries) is incomplete, without a [letter] *vav*. This teaches us that the light emanating from celestial bodies is not equal to the richness and purity of the primeval light." For greater detail one needs to refer to the more specialized works, such as the *Bahir* or *Zohar*.

The *Bahir*, which starts with a statement about a brilliantly bright (*bahir*) sky, is one of the earliest Kabbalistic works, attributed to Rabbi Nehunia ben haKana of the first century. The attenuation of the very first, amazingly bright light forms part of the discussion. The same discussion is found in many later, non-esoteric commentaries on the Torah. In all instances we are told that the initial light was bright beyond description. No physical being could withstand it. Creation of a physical universe with creatures like us could not proceed in the presence of this light, but the light is nevertheless a necessary part of the creation process. We have to start with the light. That is also what science says. The initial dense energy state was a precursor for all that followed. Science shows that the initial energy density was attenuated as it spread out into a larger and larger expanding universe. Torah also describes an attenuation through a process of contraction called *tzimtzum*. Initially there were ten vessels, or emanations, called *sefirot*, that contained the bright light. But soon seven of the vessels shattered and had to be reconstituted in a different form. Now only three, or possibly four, of the *sefirot* are accessible to us directly. The others are part of the world, but we cannot directly perceive their existence. So it is also with science. The original unified force upon which the early universe depended shattered into four forces that we perceive as separate entities. The weak and strong forces, electromagnetism and gravity. Many of the world's top scientists are trying to determine how these are, or were, interconnected.

This writer is totally unqualified to deal with anything involving the hidden knowledge of Kabbalah. There was a time when such knowledge was discussed only among those who already knew it, and it would be taught only as hints of chapter headings to those of exceptional ability and piety. Even the lesser knowledge involving basic aspects of creation, known as *maaseh breishis*, was to be discussed only with no more than two students at a time. Today these matters are much more widely known, as science has caught up with Torah. But those of us who are not among the privileged few, and that is almost everybody, really do not know what it is that we do not know. There are a large number of tantalizing hints and "coincidences" floating about. But I would not dare pretend to claim any understanding. Here is one example.

The Torah tells us that there is a force, or a field, or a presence or a whatever one may wish to call it, permeating the most fundamental and deepest part of this universe. This is an essential element of physical existence. Without it there would not be any universe as we have it. (For example, see Isaiah 6:3. We read there that "the whole world is filled with His glory," where world means universe, and not just the planet Earth.) This force field, as I will call it, is very powerful. It is more powerful than the aspects of the universe that we perceive, but it is invisible to us. We are told that initially the universe was in a state of *tohu* and *bohu*. These concepts are usually understood as confusion, chaos, emptiness and desolation, where *tohu* is chaos, and *bohu* is emptiness. But this emptiness is a false emptiness. It only seems empty to us, but it is filled with the stuff that keeps the universe going. Yes, *bohu* means empty, but it can also be read as the two words *bo* = in it, or within, and *hu* = it is (is it, in Hebrew). The underlying fabric of the universe is contained within it.

Is there a science based-analogue to this puzzling description? Yes, there is. Most people are not aware that the bright-matter we observe in the universe accounts for no more than 5% of the total content. For years scientists have searched for the remaining "dark matter" of our universe. This has been a most difficult problem because all indications are that the hidden, dark matter, can not account for more than 30% of the needed total. Recently, theories and measurements indicate that the mass density of quantum vacuum fluctuations accounts for the missing 70%. Not only that, but this energy, stored in the fabric of space, is causing the expansion of the universe to speed up and it lets us reconcile puzzling differences in age of the universe measurement disagreements. Analysis based on this idea shows a universe age of 14.2 billion years (BY). This is exactly equal to a Torah-based calculation of 14.23 BY, as discussed elsewhere in this book. Martin Rees notes on page 95 of his *Just Six Numbers* that "The US-based magazine *Science* rated this as the number-one scientific discovery of 1998 in any field of research." This recently discovered powerful, repulsive force, has three times the energy density of the matter content of the universe, but we can only get to it by indirect inference. It is invisible to us as indicated by the Torah.

Δ This assumes a static universe. See footnote to Chapter 2 for more details on the event horizon distance.

ΔΔ Alpha Centauri is actually a triple star, whose closest component, Proxima Centauri, is 4.3 light-years from Earth. I have chosen to treat Alpha Centauri as one item in this book, for the sake of simplicity.

Chapter 6

Hashem Is Here, Hashem Is There

Understanding Time

There is a well-known children's song that begins with the words, "Hashem is here, Hashem is there." *Hashem* means the name. But it is not just any name. This is The Name. The Name that we do not pronounce as written. The Name of four letters, the Tetragrammaton. Given the somewhat frivolous nature of a children's song, we would not pronounce the Name even in its secondary form. Hence *Hashem*. But everybody knows Who the Name refers to, even the children. The children sing the song to a lively tune, and they make gestures in accordance with the lyrics. Hashem is here, Hashem is there. Hashem is right, Hashem is left. The children point right, and they point left. The children point up, and they point down. They point in every way that one can, involving the six directions of our three-dimensional space. And they finish with the refrain that Hashem is everywhere.

The interesting part to me is that even the youngest children seem to comprehend the basic concept that they are singing about. Apparently, it is not counterintuitive to consider a Being extended over all of space. To be sure, most of us use inappropriate anthropomorphic analogies, such as smoke dispersing to occupy all of an available volume. Still, the idea is not uncomfortable even to children. But then, we are after all dealing with a children's song. Now, if it were a song for adults, that would be different. Indeed it would. And the first change we would need to make is to change from everywhere to everywhen. Can't find *everywhen* in your dictionary? Neither can I. There is no such word. But that is the concept that we need to deal with.

Here is how Rabbi Munk explains it on page 97 of *The Wisdom in the Hebrew Alphabet*, "To the Infinite Creator, there is neither past nor future; in the Divine conception, which is unfettered by place or time, everything is as accessible as the present." Maimonides has summarized our principles of belief based on the Torah in his commentary to *Mishnah Sanhedrin*. Among these is the understanding that Hashem is not corporeal and that He transcends all space and time. We also hold that He knows all that transpires in all time and space, including our thoughts. Nevertheless, we have free will. Maimonides, of course, did not invent these and the others of the Thirteen Principles of Faith on his own. He simply summarized in a succinct form what the Torah tells us. Indeed, we need not go very far to find the connection to all space and time, because the very name of Hashem includes this information.

The four-letter name, which in English might be written as YHVH, can be understood in many ways. One of these ways is connected to space-time by the meaning of these letters. Three of the letters form an acronym for the Hebrew words of past, present and future (*hayah*, *hoveh*, *yihyeh*). The fourth letter is the *vav*, which represents the numerical value of six in Hebrew. A symbol of the six physical directions associated with three-dimensional space is inserted between the letters representing present and future. Our present is connected through physical space to the future. Not only is all time, whether past or future, the same to Hashem, but space and time are not separate entities. These are interconnected in a single undifferentiated whole.

The Torah does not treat of time or space but of space-time. A most modern 20th century physics concept. Furthermore, these are not special or esoteric concepts. This is the normal, everyday understanding of our belief. We speak of this in study and prayer. Thus, the well-known hymn, *yigdal* (exalted), contains the line, "unbounded by time is His existence." It would seem that such ideas are too mystical to be reconciled with science. But that is not the case. Modern science is up to the task, as is exemplified by the connection of space and time into one unified space-time. Let us consider then some of the aspects of space and time, and especially time, inherent in the latest scientific investigations.

The arrow of time

We all know the difference between the past and the future. These are clearly different. We drop a plate and it breaks. We remember that the plate was dropped and that it broke, but only after it happens in the past. We do not remember the incident in the future, before it happens. We describe a time-based sequence. We claim that the plate broke because it was dropped. No one would plausibly suggest that the plate was dropped because it broke. There is a time sequence here. A cause and an effect. A past and a future. A direction, which scientists call the arrow of time. But the fundamental laws of physics that affect us directly do not require this. Gravity and electromagnetism, the forces that affect us directly, are completely time-reversible. And causality, the idea that cause must precede effect, is not an inherent requirement of the laws of the universe as now understood. Things are not as simple, or as obvious, as they might seem.

Hit a billiard ball, which hits another ball, and these move about in certain ways. A physicist analyzing the time-sequence of forces – the velocities, the momentum – will find that everything works in accordance with the known laws of the universe. Now do everything backwards, moving from the future to the past. Everything still works as well as before. There is no preferential direction of time. There is no arrow of time here. Take a super magnifying glass and observe the behavior of three molecules as these jiggle about in response to the effects of pressure and temperature. Measure and record all the movements, all the collisions and check against the known laws of physics. Everything will be in order. Now do all the calculations backwards, moving from the future to the past. Everything will still be in order. Make a recording and show the sequence to people. Show the sequence forward, and show the sequence backward. Will they know which is which? Will they say that one looks right and one looks wrong? No, everything looks right. Backward or forward, it makes no difference. The direction of time does not matter, either to the physicist making measurements or to the observer relying on experience. We can not tell the difference, because there is no difference. But we know that time does have a direction. We know that past and future are different. What is going on here?

What is going on here is that we are missing the forest by concentrating on the trees. We are looking at too fine a level. Take a box full of air molecules, millions and billions of them. Now let a tiny bit of smoke in through an opening. The smoke molecules slowly disperse through the box till everything is a uniform gray. Look at any one smoke molecule. It moves right, and it moves left. Sometimes it moves away and sometimes it moves towards the opening in the box. There is nothing wrong in the physics, and nothing strange in appearance, when this molecule moves back towards the opening in the box whence it came. But now consider all the smoke molecules. We expect these to disperse uniformly throughout the box as time passes. No one would expect the smoke to go swooshing back out of the box through the little hole on the side. Do the laws of physics permit this? Yes. Anything one molecule can do, one million molecules can also do. But the probability of this happening is, as the saying goes, astronomical. In fact, the probability is beyond astronomical. The universe just hasn't been around long enough for this to happen. In fact, a large number of "impossible" phenomena are possible. It's just that the probability is vanishingly small. But the probability of going in the other direction is very, very high. There is a preferential direction for how conditions develop over time. It lets us differentiate between past and future. We have an arrow of time.

The discussion above, about the molecules and the smoke, involves what is generally known as the *thermodynamic* arrow of time. This well known indicator of time direction is based on the second law of thermodynamics. This law states that the quantity of disorder in a closed system, as indicated by the measure of entropy, increases with time. This tells us also that uneven temperatures in a system will equalize over time to one common temperature. And that, while the total energy will not change, the amount that is accessible to do useful work will decrease in time. And so forth. The thermodynamic arrow, the scientific indicator of a preferred direction for time, and the psychological arrow, that we all experience as we remember the past but not the future, are likely the best known. But there are other direction-of-time indicators. The electromagnetic arrow of time, for instance, is based on the fact that we can send light and radio signals to the future but not the past. We see the sun as it was eight minutes previously, because it takes eight minutes for the light from the sun to reach the Earth. This is an example of a signal from the sun getting to us eight minutes in the future, after it is transmitted. The cosmic microwave background radiation, which is the remaining remnant of the initial creation, is the oldest time-based signal that we are now receiving. This signal started on its way some 15 billion years ago.

There are debates among scientists as to which "arrow" is the more fundamental. Which stands on its own, and which can be derived from another one. Currently, there is no consensus on this matter. There is general agreement, however, that the so called "CP-violation arrow" is of great importance. Standard theories require that direct transformation of energy to matter yield equal amounts of ordinary matter and antimatter. These particles are identical except for electric charge and the fact that matter and antimatter annihilate each other upon contact. Our current universe is the remnant of a tiny excess of matter over antimatter in the early universe. All the remaining particles having annihilated each other. Where did the extra matter come from? The CP-violation gives a clue. It was found that a charge-exchange = C; and a reversal of space directions, called parity = P, yields different results in the decays of a particle identified as a certain type of meson. A universe with a time arrow as it now is, will yield more matter than antimatter. This explains some, though not all[Δ], of the excess of matter over antimatter in the universe and shows a fundamental difference in result with time reversal.

It should be clear that while the basic laws of physics, such as those governing gravity effects or electromagnetic behavior, show no time-based favoritism, there are behavioral principles which in fact show a time preference. But none of this explains the behavior postulated by the principle of causality. We take it for granted that cause must precede all effects, no matter how we look at it. But we do not know where this requirement comes from. Furthermore, "precede" is not a simple concept as demonstrated by relativity and quantum mechanics. In any event, we can note that scientists hold this "obvious" matter to be quite mysterious, and without current explanation.

It is interesting that Torah says the same thing. Rabbi Dessler, whom I have quoted extensively on the nature of miracles, says, "What precisely is cause and effect? Why does the effect proceed from the cause? ... It must be admitted that we have no answer to the question of why the effect follows the cause. All we know is that this is what invariably happens." Torah has no problem with this lack of understanding of cause and effect in everyday life. There is no real cause and effect in everyday life according to Torah, because ultimately everything derives from one Cause. Science says essentially the same thing. We have a principle of causality, which states that the effect follows the cause. We know of no violation of this principle. But we do not know why things are this way.

Time travel

Most of the logical obstacles to an acceptance of "everywhen" come from an anthropomorphic-based fallacy. We look at what would be involved to have a being who is somewhat like us explore all of time. In other words, we are talking about time travel. This immediately gets us into the logical difficulty of the time-loop paradox. There are numerous paradox variations. Here is a simple version. A time traveler goes to the past and kills a grandparent in childhood. Obviously the time traveler is never born. So no one goes back in time. Therefore, the grandparent lives and has children, and the time traveler is born. Now the time traveler goes back in time and kills a grandparent. Hence, the traveler is not born and does not go back in time. We are stuck in the time loop, and apparently unable to get out. Anyone acquainted with science fiction literature will be familiar with numerous time-paradox versions, and also numerous resolutions.

One of the possible solutions is to invoke a "many universes" quantum mechanics interpretation. This interpretation of quantum mechanics behavior calls for the constant slitting into separate universes at every decision point. It is a logical consequence of the strong anthropic principle, and it eliminates many logical difficulties with the big bang theory. Why do we have this universe, which is so highly improbable? One response is to claim that it was done on purpose, as the Torah does. Another response is to claim that there are an infinite number of universes, and this just happens to be the one where we exist. This is attractive as it eliminates the need to invoke the Bible. But the Torah does not object to the possibility of multiple universes. Multiple universes are noted in Talmud Tractate

Avoda Zorah, 3b. This matter is further elaborated on by Rabbi Judah ben Barzilai of 12th century Barcelona. He not only discusses multiple universes, but also speculates on whether there are living, intelligent beings in any of the universes other than our own. A scientific theory involving multiple universes does not falsify the Torah. Furthermore, it provides a science-based explanation for "everywhen." Here, everywhen would involve interaction with this universe from outside the universe. This is permitted in a multi-universe system, especially if this universe is an outgrowth from a more fundamental, a higher-level universe.

Another "natural," or scientific, example of everywhen involves the velocity of light. Einstein's special theory of Relativity shows that there is an absolutely maximum velocity that cannot be exceeded. It is generally referred to as the speed of light, and designated by the symbol "c." The word "light" is a general designation, and does not refer specifically to visible light. Any sort of electromagnetic radiation will qualify. The speed limit is c = 299,729 kilometers per second in a vacuum. This limit cannot be exceeded, no matter what you do, or where you are, or how you move, or how you measure or what is moving. Furthermore, material particles, which have mass, cannot move that fast because the mass increases the closer the velocity gets to the speed of light. The mass becomes infinite at the speed of light, and that makes it impossible to push along. But the massless photon, associated with a light wave, has no such difficulty. This is something special for the photon that would not apply to a mass-bearing particle, like the electron. But there is something else special about the photon. The photon does not experience time duration. The photon exists in a perpetual "now." There is no past, and there is no future, there is only now.

The light from our sun takes eight minutes to get to us. The light from Alpha Centauri, our closest stellar neighbor, takes 4.3 years to get to us. The light from our closest galactic neighbor, Andromeda, takes two million years to get to us. The light remnant from the big bang beginning of the universe took 15 billion years to get to us. But that is not the time passage experienced by the photons. All of these photons experienced the same amount of time duration, namely zero. No, I am certainly not suggesting that Hashem consists of photons, though light is one of the analogies used in Torah. Nevertheless, it helps to understand that a perpetual "now," bizarre as it seems, is not a concept contrary to scientific thinking.

There is yet a third science-based understanding for being outside of time, or encompassing all time. My scientific specialty is a field called spectrum analysis. We compute the spectrum by solving a mathematical equation called the Fourier Transform. Fourier mathematics-based procedures are quite common today. Nobody thinks much about it. We just solve the equations and go on with our business. But it was not always so. Fourier could not convince anybody that his ideas had merit when these were first proposed in 1807. It was not till 1822, when rigorous proofs were provided, that his ideas were finally accepted. And what is the essence of the strange idea that Fourier was proposing? That time and something called the frequency spectrum are completely equivalent.

These are known as different domains, today. Fourier mathematics provides a transformation between the time domain and the frequency spectrum domain. The important point is that these domains are completely equivalent. The same information is contained in both domains. Both domains extend from minus to plus infinity. That is, the time domain extends over all time, and the frequency domain extends over all frequency. To be sure, infinity is a difficult place to get to, whether in time or in frequency. Hence, real measurements involve approximations. But real calculations using mathematical representations do not need to make approximations. We solve an integral from minus to plus infinity, and we get an answer. This answer contains all the information for all of time. But it is not expressed in units of time. Time, all of it, past, present and future are interconnected and can be combined into one concise expression which contains everything. People not used to this have a difficult time with the concept. This is not unreasonable. We know that the best mathematicians of the time at first thought that Fourier had lost his reason. But the mathematics is sound, and the procedure works.

Finally, here is a fourth science-based procedure that gets us outside the conventional understanding of time. Gravity controls behavior at large spaces and for large objects, while quantum mechanical effects describe behavior for small distances and for small objects. Attempts have been made to find a single theory that incorporates both these effects. No such finished theory exists at this time. But certain features have been worked out. Among the accepted procedure is this one: to find the location of a particle, we use a "summation of all possible histories" process developed by physicist Richard Feynman. A particle has a dual nature in quantum mechanics. It is both an extended wave and a single location object simultaneously. Feynman suggested that the wave version of a particle moves between two points by every possible path that a wave can follow. We find the start and end locations of the particle by adding all of these waves which then collapse, or cancel out, everywhere except at the actual particle locations. This is called the "sum over histories" process. But there is a problem. The mathematics does not work unless we use so called "imaginary time."

"Imaginary" is a mathematical term and does not have the common meaning of this word. It relates to the square root of minus one. Real time introduces singularities, sort of infinities, which the mathematics cannot handle. Using imaginary time makes for a four-dimensional Euclidean space-time, where space and time are indistinguishable from each other. Stephen Hawking asks in his *A Brief History of Time*, which is reality and which is imagination. Is it possible that the universe actually operates on the principle of what we call imaginary time, while that which we call real time is only an illusion? After all, imaginary time is more logical and reasonable from a purely mathematics-procedure point of view. There is no separate space or time when viewed in imaginary time. There is no past and there is no future. Time is the same as space, and space is the same as time, and there is no restriction as to transitions or locations within this space-time. All space and all time is open before us.

The above shows that there are science-based methodologies that can be used to explain how all time can be encompassed. But this is actually not needed. This is because we are dealing with a Being who is outside of space, outside of time, outside of the matter and energy of this universe, not subject to the laws of physics, incorporeal and in all ways totally different from anything that we can conceive of. We are not talking about a change in degree, involving some unimaginably advanced alien intelligence. We are talking about a change in kind. An analogy might be the difference between discussing a googolplex – a huge number that we cannot really comprehend, but still one which, because it is in the realm of ordinary numbers, we can still try to understand by extending the idea of a million or a billion – and discussing transfinite numbers. In the latter case, the mathematics has to change. Numerous works, including Maimonides' *Guide for the Perplexed*, Luzzatto's *The Way of G-d* and many others, offer us a mere glimpse of the hidden reality. This author has nothing new to contribute at such a level of scholarship.

Cause, effect and free will

We are told in chapter 3 of *Mishna Pirkei Avot* that "Everything is foreseen, yet the freedom of choice is given." How is such a combination possible? How can the future be known, or foreseen, if we can choose to act in any manner we please? This is not the only apparent disconnect between science and Torah. As I noted earlier, some people have difficulty with evolution. Some people have a problem reconciling science and Torah perspectives on the physical universe. Many people are terribly troubled by good and evil, unable to find a satisfactory answer to the question why the innocent and righteous suffer while the wicked prosper. Different people have differing perspectives.

In my opinion, the most difficult discussion revolves around causality and free will. Creation and the evolution of both the universe and living beings is a most serious matter. But this writer believes that science and Torah are in remarkable agreement on these, as discussed in other essays in this book. Good and evil, prosperity and suffering, who can be immune to these? Our Sages have provided powerful answers to these questions. Those who believe in Torah will accept the answers, and those who do not believe in Torah might not accept the answers. Either way, science has nothing to do with it. But cause and effect is a scientific concept. Encompassing all of time so that past, present and future are one and the same has scientific implications. We appear to have a genuine conflict here between science and Torah. And the resolution is not at all simple. Possibly, the reason why this matter appears to be so difficult is that we are not dealing with one issue. Separating the matter into separate, independent parts will make this seeming conflict much more tractable.

One of the traps into which we are all prone to fall has already been discussed. It is the nature of the Being Who foresees the "everything" in the quotation with which I opened this section. This is not a supercomputer which has been programmed to forecast the future based on current events. This is not an extraterrestrial alien equipped with a superior technology. This is not any sort of being that is of this universe. This is what the Torah tells us. The Torah gives us the same answers today that it gave last year, a century ago and a millennium ago. The Torah has not changed. Science had very little to say about this in prior times. Today science has progressed to the point where it does have something to say on this matter, as discussed at the outset of this essay. The best that science can say right now is that the existence of such a Being is apparently not impossible. And even if science said that it was impossible, it still would not falsify Torah because this Being exists outside the laws of this universe. Thus, quoting from Rabbi Dessler in *Strive for Truth*, "Foreknowledge of a person's free choice is possible only to the mind of Hashem, who is above time and the nature of whose knowing is completely different from the knowing we experience within creation."

The other part of the equation involves us, whose freedom to choose is in question and whose actions are foreseen. There was a time in science, soon after Newton published his insights, when determinism held sway. Thus, Pierre de Laplace assured the world, and the world believed, that the future could be predicted with absolute accuracy. Not only inanimate particles, but we too, being subject to the same fundamental laws of the universe, were merely acting out a drama to a previously determined script. To be sure, it would take a much more advanced technology than we had at the time, or than we now have to do it. But nevertheless, free will would appear to be an illusion, if only in principle. Torah has always disagreed with this view. Now science also disagrees. We know now that the future cannot be absolutely predicted, and this is not due simply to inadequacies in our technology.

Peculiarities associated with Relativity, such as cause and effect reversal due to frame of reference issues, get in the way of accurate predictions. We know today that the behavior of a large aggregate of particles needs to be described by probability distributions, even though each individual particle interaction is perfectly describable. This is what gives rise to the thermodynamic arrow of time. We know that time is anything but a simple concept involving a linear sequence or progression of events or changes. Simultaneity of events, rate of change of time, identification of cause and effect, the reality of linear real time and various other factors make time one of the most complicated and difficult physical concepts to describe.

To predict, indeed to determine, the future requires that there be both a past and a future. We have chaos theory involving systems of virtually infinite sensitivity to infinitesimal changes in initial conditions. How does one calculate a future result for such a case? Finally, we have quantum mechanics. Matter and energy appearing out of nowhere in the vacuum. Indeed, even space and time itself being created out of quantum fluctuations. And who has not heard of the uncertainty principle? Just that alone ensures that absolute determinism is dead.

Science does permit free will, because all results are not strictly predetermined. The processes that achieve this are most subtle, which makes some people unhappy. What, after all, do quantum mechanical effects on the position of an electron have to do with determinism on the scale on which people interact? I don't know. But that is actually what the Torah requires. We are told that the universe is substantially deterministic precisely so as to permit free will. This is because we need to decide on the basis of expected results. A chaotic universe, where the same cause resulted in totally different effects, would destroy our ability to decide. What good would free will do for us under such conditions? What is needed, according to Torah, is that the world should appear deterministic but in actuality not be fully deterministic.

Science permits free will, and science also provides tantalizing hints about the possibilities of encompassing all space and all time in one integrated whole. It is possible to foresee the future. It is possible for us to choose and act. One does not preclude the other. Yet, as creatures living in real space and time, it is not possible for us to experience such a combination. We cannot jump into an imaginary time-based universe, for example. And we cannot truly understand how it works. But science does not say no, it cannot be. Neither does science say that this universe cannot be, or is not, maintained and renewed on a continual basis. Science does not say that our lives and our futures are not affected or acted upon, in spite of our free choice to act. Both are logically possible. Yes, there are profound theological, philosophical and logical issues here. But there is no fundamental conflict between science and Torah. That is enough to keep this writer satisfied.

Δ Late breaking news, in 2001, affects a number of matters discussed in this book, including the mystery of the excess of matter over antimatter. We noted that the CP violation for the kaon (K meson) accounts for some, though not all the difference between matter and antimatter. An announcement on February 12, 2001 relates to new evidence, obtained at both the SLAC (Stanford Linear Accelerator) and KEK (Accelerator Research Organization in Japan), that shows to a 95% certainty, that decay parity is also violated for the more massive B meson. This would pretty much account for all the difference between matter and antimatter, assuming that the preliminary results are verified to a higher level of certainty.

Other late-breaking news is equally exciting. On February 9, 2001 the Brookhaven National Laboratory announced results that point to verification of a theory called "supersymmerty," which among other consequences, points to a Higgs boson at a 130 GeV mass. This is in line with recent tantalizing hints from CERN in Switzerland that the Higgs particle might be detected at a level of just over 115 GeV. All of these discoveries are in line with what the Torah tells us about the universe. The reader will find more details on these discoveries in the March, 2001 issue of the *IEEE SPECTRUM*. ***New Light on How, and Why, Matter Materializes***.

Chapter 7

The Heavenly Time Machine

Age of the Universe

We've discussed some aspects of time travel elsewhere. It was made clear that the idea of time travel applies to created beings like us, but not to the Master of the World, who exists outside of time and space. The previous discussion about time travel involves a device, a machine, that would transfer one to a different instant of time. There is also a different concept of time machine, the one we have in this essay. This "machine" does not move within time, but rather changes the rate at which time moves. Is it possible to reconcile the Torah-based six days of creation and nearly 6000 years of existence with 15 billion years based on scientific theory? Surprisingly, the answer is yes. And it is not terribly difficult either.

Is Torah science?

We have a science-based calculation which transforms 15 billion years (BY) into six days through a time dilation process. This calculation is not dependent on Torah. It is strictly a scientific procedure. A scientist has a basis upon which to judge the validity of results. We also have a number of Torah-based calculations that lead to an age of the universe of between 14 and 16 BY. These calculations are not connected to science but are based only on Torah. Hence, a scientist is entitled to ask for some validation, or explanation, of the process. Unfortunately, this writer is not competent to explain Torah at such a level. I know what certain people have to say on this subject, what verses they quote and what results they obtain. But why it fundamentally works that way is at an entirely different level of inquiry. Someone who has some personal experience of the depth of wisdom exhibited by these people would accept that they knew things that we do not know. But not everyone has this experience. What does one say to a Feynman who, as described in the first chapter, thinks that the Talmud is an interesting book, which is characterized by extra large pages and a medieval form of reasoning? All I can do is try to show that these people deserve more credit than that, even when judging on a purely scientific basis.

Neither the Torah, nor the commentaries on the Torah, are books of science. But these do contain science-based information. There are, for example, significant sections on mathematics, such as how to accurately calculate the value of pi. Another frequent topic is in the field of anatomy and medicine. A third common field involves astronomy. Astronomy is the closest to the content of this chapter, so let us take an example from astronomy. We have a statement on the length of the lunar month on page 25a in Tractate *Rosh Hashanah*. This would be the cycle of the phases of the moon, otherwise known as the synodic month. Rabban Gamliel says that he has it from the traditions of his grandfather's house that the length of the month is 29 days plus half a day plus two-thirds of an hour plus 73 parts of an hour. An hour is divided into 1080 parts in Talmudic measure. Addition of these numbers yields a lunar month that is 29.530594 days long. Current measurements, as reported in the Encyclopedia Britannica, show a mean time period of 29.530588 days. The difference is 0.000006 days, which is a touch over half a second. Roughly two thousand years ago, somebody established the length of the lunar month to within 0.5 seconds of the best current value. Admittedly, the result is not as precise as it seems, because the length of the month has changed over time. Nevertheless, we have here a remarkable coincidence, or what else can it be? We are not told how the result was obtained, so we are free to speculate. There are only two logical possibilities left if we are to dismiss the idea of a lucky guess. Either the answer is hidden within the Torah, or these were first-rate scientists who made sophisticated calculations based on results from careful measurements. Indeed, there is speculation that the result was derived from calculations based on the measurement of elapsed time between eclipses. Whatever the procedure, it would seem that these people are not the medieval reasoning simpletons that Feynman so readily dismissed in Chapter one.

How old is the universe?

One can conjecture a year more or a year less, but the plain meaning of the Torah is that the universe is less than 6000 years old at this time. Yet serious commentary claims a range of between 14 and 16 BY for the Torah. There was nothing in the conventional wisdom or science of the day to point to millions, let alone, billions of years. The Torah Sages of those days were not completely ignorant of science and mathematics, as discussed with respect to the lunar month. What then would compel these people to propose ridiculous-seeming numbers like 15 BY? There is only one logical conclusion. These people made the claims that they did, because they were convinced that the evidence was in accord with a deeper meaning of Torah. Today these numbers are right at home. Where do these numbers come from, and how do we reconcile these to 6000 years?

The reconciliation is actually fairly simple, once we understand that the first six days of creation are not counted in the age of the universe. The Torah starts the count from the beginning of mankind and not from the beginning of space and time. Therefore, the question is not how 6000 years become 15 billion years, but rather how six days become 15 billion years. Modern science has an answer for that. Like all clever ideas, it seems quite simple once it is explained. The first explanation that this writer knows about appears in *The Science of G-d* by Gerald Schroeder (Simon and Schuster, 1997). The universe was once small. Now it is large. The universe was once dense, now it is fairly empty. The universe was once hot, now it is cold. The universe was once young. Now it is not so young. The early universe was different in many ways from the way it is today. Many parameters were different, including the flow of time.

Most people are familiar with the slowing of time due to high relative motion, known as time dilation. It is a direct consequence of the special theory of relativity which in turn is based on observational evidence that the speed of light in free space has the same value for all observers regardless of their motion relative to each other. This is totally contrary to our experience and to common sense. You stand near the railroad tracks and watch a train pass by at 25 miles per hour. You measure the speed of motion of a passenger walking at a brisk 5 miles per hour inside the moving train. What do you get? You get 30 miles per hour if the person is walking towards the front of the train. And you get 20 miles per hour if the person is walking in the opposite direction. The speeds of motion add linearly. We either get 25 + 5 = 30, or we get 25 – 5 = 20, depending on whether the walk is with the motion of the train or opposed to it. Now suppose we have a super train that runs at one quarter the speed of light. A flashlight is turned on by a passenger in the train, pointing to the front of the train. People on the train will measure the speed of the light beam from the flashlight at 299,729 kilometers per second, normally designated as "c" or the speed of light. People not on the train will find that the train is moving at 25% of the speed of light, but the flashlight beam is still measured at only 100% of the speed of light, and not at 125%. And if another train passes the first one in the opposite direction and also moving at a quarter the speed of light, the flashlight beam will still register at 100% of the speed of light.

Well, if the speed of light is constant then something else has to change. And it does. Objects get shorter in the direction of movement. Mass gets bigger. And time interval is reduced. The time dilation effect is not just a theory. It has been verified by experiment. This strange result is illustrated by the well known twin paradox. One twin travels to a distant star using a very fast space ship that moves at a substantial portion of the speed of light, while the other twin stays at home. The twin that travels ages less than the one that does not. The closer we get to the speed of light, the slower the rate at which the internal clock changes. There is no limit to this time dilation effect. A day can be a week, a year, a million years or a billion years. It all depends on how fast we move. Indeed, the passage of time stops completely when the velocity, "v," reaches the speed of light, "c." Of course, we can never get there because the mass of the traveler increases without bound. But we can get as close as you like, and the time dilation can be as large as you like.

Another time dilation effect is provided by the more advanced version of relativity, known as the General Relativity theory. Here it is shown that a high gravitational field is equivalent to a high acceleration. You stand in an elevator moving up, and you feel a force pressing at your feet. You stand on the earth, and you feel a force pressing at your feet. It is all the same, and the result is the same. You will age less on Jupiter, where the gravitational attraction is greater, than you will on the earth. You will age more quickly on the moon, where the pull of gravity is less, than you will on the earth. Careful measurement using a high-accuracy "atomic" clock shows that time moves slower on Earth than in space where the force of Earth's gravity is not a factor. Not only is the change in timing in the right direction, but the amount of change is exactly as predicted. There is no question about the correctness of the time dilation due to gravitational effects. Want to really slow things down? Go to a neutron star where the force of gravity is not just big, but super huge. Unfortunately, you would not be around to enjoy this because you would immediately be squashed into the size of a microscopic dot. But we can still talk about this, even if we cannot actually do it.

Many people are familiar with the above two time dilation effects. Dr. Schroeder introduces us to a third version which is directly connected to the six days of creation and the current age of the universe. What happens as we sit here on earth and the sun emits some light photons? We detect this emission about eight minutes later because that is how long it takes the light to get to us from the distant sun. We can say that the sun is eight light minutes distant from us now. We can also say that the sun was eight light minutes distant from us previously. Our distance to the sun has not really changed during the transit time of the light. Now consider the cosmic background radiation (CBR). According to our time measure, it is the remnant of a very hot universe some 15 billion years ago. But the object that emitted this "light," if we can call the whole universe an object, was not 15 billion years distant from us when this so called light was emitted. This is because the universe is expanding. It started when the whole universe was smaller than a grain of sand. The Torah, as explained by Nachmanides, uses the analogy of a mustard seed. Indeed, we are told that this creation "was like a very small point having no substance" (see page 25 of the Chavel translation of Nachmanides on Genesis). Not only is everything moving apart, but this is not an ordinary movement. This is not the case of an independent object moving from one location to another location in space. This movement is due to the fact that space itself is stretching, as new space is being created. This light once filled the whole universe, and it still fills the whole universe, but the universe is immensely larger. Velocity, time and distance must maintain their relationship. We know that the distance changed, so what happens to time if the velocity does not change?

Schroeder provides a fairly simple formula for the time dilation [Δ] due to the creation, or stretching, of space as the universe expands. One might think that any such calculation would be inaccurate to the level of uselessness because we need to know the rates of expansion and relative sizes of the universe. Fortunately, that is not necessary because the temperature of the CBR changed in step with the expansion of the universe. All we need to know is the temperature and we have an answer. The current temperature is known very accurately at 2.73 degrees. But where do we start? Nachmanides, who combined Kabbalistic teachings with his commentary, tells us that the proper point is where ordinary matter, what a physicist would call hadrons, came into being. We can understand this in two ways.

One is when the temperature dropped to the point that the fundamental particles from which matter is composed, called quarks, could no longer exist on their own. This is identified as a phase transition from a quark-gluon plasma to hadron matter. Ordinary hadron matter is composed of quarks held together by gluons. This temperature is expected to be around 10^{13} degrees. But we do not really know because the phase transition probably went through a supercooled state where the temperature dropped quite a bit before the actual phase change. An analogy would be to cool very pure water below its freezing temperature and have the water stay liquid as long as it is not disturbed in some way. Serious experiments are in progress to determine the phase transition temperature.

An alternate high temperature point is to use the temperature where the radiation photon energy equals the rest mass energy of the neutron or proton. This is the temperature at which photons transitioned to ordinary matter in the early hot universe. This number can be accurately calculated at 10.9×10^{12} degrees. The reader will find the result on page 156 of Weinberg's book *The First Three Minutes*. This is the value used by Schroeder. Plug into the equation the starting temperature and the ending temperature and presto – the first six days of the universe as viewed from their frame of reference are as 15.75 BY from the current frame of reference. This number is remarkably in line with one of the suggestions from Torah commentary. But my personal preference is 14.2 BY, for reasons that will become apparent below. This would require that the starting temperature be 9.7×10^{12} degrees, which is within the range of a supercooled phase transition. It is hoped that high-energy particle collider experiments will soon determine the temperature value of this phase transition, and then we will know what temperature to use. In the meantime, we can only marvel at the fact that six days comes to within a few percent of the best estimate for the age of the universe that science now has to offer.

The first six days

The first six days, involved in the creation, are not counted as a part of the current age of the world. Hence it is possible, and permissible, to use these separately and outside the current age of near 6000 years. This brings us to a question. Are these 24-hour-type days, or are these indefinite periods of time that can take on any value, even billions of years? The answer that it is both, creates a paradox. Science, through the agency of time dilation, has resolved the paradox. And while these days last both 24 hours and simultaneously billions of years, they are not of indefinite time duration. A deep understanding of Torah provides a specific time period which just happens to fit current science-based results. The reader is welcome to compute the probability that it is all just part of a coincidence.

There are numerous commentaries, derived from Torah, related to the initial creation process. There are discussions related to the use of the alphabet. There is commentary related to statements or sayings. And there is commentary related to the use of the Torah as a template or precursor of the creation. All of these are understood as not being literal explanations of what happened. No one claims that there were literal statements uttered, with sounds that could be heard, as the drivers of creation. These are simply intended to teach us some deep truths in a language and with analogies that people can relate to. We know that in reality no human being can truly fathom what actually took place.

One such understanding is that creation was initiated by the establishment of the Torah 1000 generations before the Torah was handed over to mankind. One will find, among others, a discussion on such matters in *Midrash Rabba: Breishit, Avot D'Rabbi Nathan*, and a direct reference to 1000 generations in Psalms 105:8. We know, by counting generations in the Torah, that the Torah was given to us after 26 generations starting with Adam and Eve. Hence, there are 974 generations missing. These 974 generations from a time before the first human person, Adam, are noted in Talmud *Chagigah* 13b. There are various traditions as to how long a generation lasts. Here we use 40 years. We also have a well-known tradition, based on Psalms 90:4, that a heavenly day represents 1000 earthly years. The calculation of the time span represented by the 974 hidden generations is now a simple product of the terms, including a 365.25 day year. Thus, 974 x 40 x 365.25 x 1000 = 14.23 BY.

A totally different procedure, which also leads to similar results, is exemplified by the understanding that the universe operates in cycles, which some people call cosmic jubilees. The best known presentations on this theme will be found in the writings of Rabbi Bahya ben Asher (14th century Spain), and Rabbi Isaac ben Samuel (14th century Akko). These more modern and more understandable presentations are based on the classic Kabbalistic source, *Sefer ha-Temunah*. Some hold that two cycles have passed. Some hold that four cycles have passed. Here I use the understanding of Rabbi Isaac that six cycles of 7000 years each have passed, and we are now near year 6000 of a seventh cycle. The previous cycles are not counted as ordinary earthly years, hence each day represents 1000 years. The simple result is the product of 7000 x 6 x 365.25 x 1000 = 15.3 BY, (see Kaplan's – *Immortality, Resurrection and the Age of the Universe*, for more details).

The same cosmic jubilee procedures also deal with the end of the universe, the time when all will revert back to the chaos and desolation of *tohu* and *bohu*. There are multiple results showing times of 18,000 times, 50,000 times and 350,000 times the current age of the universe. Astrophysicist Freeman Dyson has published calculations in the *Modern Review Physics* (1979) showing approximate times when unpleasant things start to happen in a continually expanding universe. Similar information appeared in the July, 1987, issue of *Discover* magazine. We find that fuel for burning stars disappears at about 7000 times the current age of the universe, and planets become detached from their burned out stars at about 65,000 times the current age of the universe. So which prediction is the more accurate? The number 7000 (science), 18,000 (Torah), 50,000 (Torah) or 65,000 (science)? We could wait to see what happens. Or we can decide that on this scale it makes no difference. Torah and science are in general agreement on the age at the end of the universe. How can that be, and what does it mean?

There are yet other tempting hints about the age of the universe in Torah if one is willing to choose a friendly interpretation. Consider the following from page 13a of Tractate *Chagigah*. The distances between constituent parts of the heavens, that is the universe, are provided in units of time. The total comes to 90,000 years. Hence the radius, representing the distance that we could go in any direction, would be 45,000 years. This is normally taken as the distance that a person could walk in this time period. The 90,000 years total comes to nearly one billion miles, using a modern measure of distance. The result represents a 30-miles-per-day rate. But one need not take this simple approach. Rabbeinu Chananel explains that the distances are actually part of esoteric and hidden knowledge because the calculation is ascribed to Rabban Yochanan ben Zakkai, possibly the foremost Kabbalist of all time. I took Rabbeinu Chananel at his word and performed a very simple calculation, using parts of the hidden knowledge that have since been revealed. We understand from the previous discussion how each day represents 1000 years. Thus, the 45,000 years represent 45,000 x 365.25 x 1000 = 16.4 BY. Now look at the following three numbers: The last calculation is 16.4 BY, the previous calculation shows 15.3 BY, and the time dilation calculation from Dr. Schroeder falls nicely in the middle at 15.75 BY.

Finally consider the following correlation between science and Torah. One of the difficulties that cosmologists have had is that different measurement procedures show different ages for the universe. The currently accepted numbers are between 12 and 16 BY. But some procedures show a lower limit of less than 8 BY. Something is clearly wrong here. Another puzzle revolves around the "missing" matter. Currently, the big bang expanding universe theory includes a process called "inflation." This process explains some otherwise puzzling aspects of the universe, such as why it was so uniform in its early stages, as determined from the cosmic background radiation behavior. This "inflation" feature also makes some predictions. One is that a parameter called the mass-energy density be exactly equal to one for the universe. But the mass we observe is way too little. Hence the puzzle of the missing matter, or unseen dark matter. But the most that current understandings will yield is a value of about 0.3, including the most optimistic estimates of the value represented by the as-yet undiscovered "dark matter." What is going on here? Is it possible that there is much more dark matter around than any currently understood mechanism can supply? Is it possible that inflation theory is wrong and the mass-energy density is much less than one? How do we reconcile these discrepancies?

In the late 1990s various astrophysicists were looking to perfect an accurate distance measurement procedure involving the very bright explosion known as a Type Ia supernova. It did not seem to work originally. But later the results were found to be perfectly correlated, after the addition of a brightness correction factor derived from the rate of change of brightness with time. Suddenly a number of things fell into place. But the result also created a brand new puzzle. It was found that the expansion of the universe is not slowing down; rather it is speeding up. The matter was reported in the scientific literature in "Observational Evidence from Supernovae for an Accelerating Universe and a Cosmological Constant," Adam G. Riess, et. al., *Astronomical Journal*, LANL Archive (May 15, 1998). Martin Rees reports on this matter on page 95 of *Just Six Numbers* under the heading "An accelerating Universe?" Even the popular press has picked up the story. My local newspaper had a headline, "Team Finds Universe Is Flooring Accelerator."

All right, so the expansion of the universe is speeding up and not slowing down. As one of the earlier chapters in this book might say, so what, and who cares? The answer is that the age of the universe discrepancies disappear once the change in recession velocity is taken into account. The result yields an age of 14.2±1.5 BY, as compared to a Torah-based calculation of 14.23 BY. This is why I would like to have the quark-to-hadron matter phase transition temperature come out at 9.7×10^{12} degrees. It would show an exact correlation between 14.2 BY and six days. Another result shows that the mass-energy of the repulsive vacuum is 0.72, while that of all matter, including dark matter, is 0.24. The total, at 0.96, is very close to the exact 1.0 called for by big bang inflation. And what does the Torah say about that? That the universe is continually being maintained and renewed by a force invisible to our direct observation. Ultimately, as is observed in Isaiah 6:3, "the whole world (universe) is filled with His glory." Where is the conflict between science and Torah?

Δ I have used the phrase "time dilation" because it is the only name I know for something like the effect I have tried to describe, even though it is not scientifically correct in this context. Relativistic time dilation is not just an equivalence calculation, but an indication of an actual change in the rate of change of time. Not so for the Schroeder formula. I am not suggesting that the rate of time flow was different, say 10,000 years ago, before the current age of the universe according to Torah, given as less than 6000 years. Rather there is an equivalence between six days and 15 BY, depending on whether we look at time starting now and going backwards, or we start at the big bang and look forward to our time.

Chapter 8

Now You See It, Now You Don't

The Nature of Quantum Mechanics

Anybody who has any familiarity with quantum mechanics (QM) knows that the stuff is weird. Niels Bohr, who is credited with leading the development of a unified theory, claimed that QM is "shocking" to anybody who tries to understand it. Most scientists agreed that, yes, it was shocking, but that is how the world apparently works. Some, including Albert Einstein, refused altogether to accept the implications of this theory, on the grounds that "G-d does not play dice." The Torah agrees with Einstein about the dice. But Torah also agrees with Bohr that this is how the world works. To us, QM is like the random results from playing dice. The neat Newtonian deterministic world is gone. But that is only on our side of the universe, and with our limitations. Einstein, one of the greatest geniuses that ever lived, was still only a very smart person. A hundred Einsteins, a thousand Einsteins or a googol of Einsteins would still be only an extension in degree of the limitations of human possibility. But G-d is a change in kind. Our limitations do not hold. Let us take a closer look at this matter of human potential and human limitations, and how these are connected to Torah, dice and quantum mechanics.

Torah, dice and free will

This material would fit equally well in the section on free will of the *Hashem Is Here* essay. The interested reader may wish to take a look at that essay for complementary material.

In *A Brief History of Time*, Stephen Hawking speculates whether what we perceive of the world, and especially time duration, is not an illusion, while the underlying reality is hidden from us. Torah says it explicitly. We have quoted extensively from Rabbi Dessler's work, *Strive for Truth*. Here is some more. "Nature has no objective existence; it is merely an illusion which gives man a choice to exercise his free will … What is real is the will of Hashem and nothing else." Why then do we have a natural order with predictable cause and effect? So that people will have a stable environment in which to exercise their free will. But "In the case where the event is hidden from human eyes … there can be no test for our free will and there is therefore no need for 'natural' cause and effect." The Torah specifically gives us a choice. Thus, "I have put before you life and death, the blessing and the curse; choose life …" (Deuteronomy 30:19). The conditions in which we find ourselves are not chosen by us – "I have put before you …" – but how we react is up to us – "choose." Referencing the Gaon of Vilna, Rabbi Dessler states that we not only can "choose," but we also have been given "the capability to realize that will." We have both abilities, to choose and also to act to accomplish the choice.

The Torah absolutely insists that we have free will. The Torah also insists that Hashem knows all, foresees all and can do or cause any result whatever. The resolution to the paradox is a world of apparent causality which we can apparently control by exercise of our free will, while at the same time the underlying reality is a world that we cannot fully predict or control. Both the apparent causality and resulting free will and the underlying reality, which is not under our control, are real and necessary. Says Rabbi Dessler, "Someone who has no idea of the underlying purpose which these two different modes of perception serve, will see a conflict." But there is no conflict between the actual control of the world by Hashem and the causality that we perceive. This is because "The perception of causality (not as an absolute, but within the bounds we have described) is also necessary so that we may discern the hand of G-d behind the causes." Torah is fully in line with quantum mechanics. One can even see, after the fact, how Torah predicts, or requires, QM effects. We need causality on our scale of existence so as to have a predictably stable environment in which to exercise free will. At the same time we need to understand that this causality is not "absolute," so that "we may discern the hand of G-d behind the causes."

The gift of free will, to choose and to act, was given to us. The Torah tells us in Genesis 1:26, "Let us make Man in Our [*tzelem*] image, after Our [*demut*] likeness ...." The meaning and implications of these words, *tzelem* and *demut*, is so important that this is the first item that Maimonides addresses in the first chapter in his *Guide for the Perplexed*. He makes it clear that the "image" of *tzelem* and "likeness" of *demut* is not related to physical appearance or any sort of body. It is because of the "intellect with which man has been endowed, [that] he is said to have been made in the form and likeness of the Almighty." We have many attributes which no other creature has. Among these is the ability to understand and affect the apparently deterministic aspects of the world; to perceive and know of the existence of the apparently random aspects of the world; and to perceive and understand that there is a Power that ultimately controls both of these "natural" type phenomena. Also there is the freedom to choose what to believe; the freedom to choose what to do about our beliefs; and ultimately the ability to act upon our freely chosen intentions. For a long time, we thought that our powers were unlimited. We thought that given time and effort we could, in principle, become full masters of this world. That is no longer the case. We have found that there are certain aspects of the world that are forever closed to us. We have tremendous potential, but ultimately it is only a likeness and an image. *Tzel*, as in *tzelem*, is a Hebrew word for shadow. "What is real is the will of Hashem." All we have is but a shadow of the real thing. This is one of the lessons that comes to us from quantum mechanics.

The uncertain world

The nature of light, whether wave or particle, was in doubt for a while. Newton championed the particle hypothesis, while his contemporary, Christian Huygens, argued in favor of waves. Ultimately Thomas Young seemed to settle the matter when he demonstrated the diffraction of light in 1801. Later, the wave nature of light was definitely secured when it was shown that light is an electromagnetic wave in line with Maxwell's theory. But the nature of light is not that simple. Max Plank's work to account for the behavior of very hot objects, followed by Einstein's work to explain the photoelectric effect, showed that light was simultaneously both a wave and a particle. It was shown that light waves are emitted in tiny bursts of energy called *quanta*. Thus was born the quantum theory of light. It is a strange idea, involving a dual nature that had not been conceived before. But it is not terribly counterintuitive. After all, a wave can be sent out in spurts. This may not be initially obvious, but it is not shocking when explained. Then in 1924 came the de Broglie hypothesis about matter waves, and things became decidedly "shocking," as Bohr was later to characterize QM.

We have no direct experience of light waves or light quanta. Somehow the light, and all other electromagnetic radiation, is granular in packets of energy but also spread out in waves. We can't visualize this dual nature, but we could never really visualize the nature of light, no matter how it might be described. That is not the case with solid matter. We have mountains, boulders, pebbles and grains of sand. These are solid clumps of stuff that we can hold and see, and weigh and measure. We can progress to smaller and smaller objects, to the point that we cannot see them directly. But we can easily visualize these as solid, individual items. QM says no, these also are waves. The object is spread out and has no localized existence. In fact, there is nothing in existence anywhere except a probability wave. It is only when the probability wave is collapsed by observation or measurement that we get a defined object. Does the object exist in the absence of an observer? Some say yes, and some say no. This certainly appears to apply to the very small, say an electron. But theory says that this also applies to the large. Even to the whole universe. Who then collapsed the wave function to bring the universe into existence? Who or what is an acceptable observer? Does it take a conscious mind, or will any living entity do it? How about a highly advanced computer? There is a long list of such questions, and we can easily digress into speculative flights of fancy. Let us then stop and get back to reality, or whatever passes for reality in the shocking, dice playing, world of QM.

QM posits that matter can spontaneously appear out of nowhere and disappear. Indeed, it is proposed that the whole universe is a result of a QM vacuum fluctuation. Electrons come and go, disappear and reappear in a different location. The universe is in a constant state of disappearance and renewal. As we are told, "He renews daily, perpetually, the work of creation." QM states that it is impossible to determine certain pairs of properties of a particle simultaneously, such as location and velocity (actually momentum). This is the often referenced Heisenberg uncertainty principle. This uncertainty is an intrinsic part of the universe and not a consequence of the crude state of our measuring technology. There are processes that we can deal with only on an aggregate basis, but not individually. We know the half-life of radioactive materials, and can predict how long it will take some percentage of the atoms to change. But we have no means to determine when some particular atom will change. We get a ghost-like probability wave any time there is a choice to be made, and the reality of the final result does not appear till the probability wave is collapsed by an outside observation.

Is an electron a wave or a particle? Experiments indicate that it is both till a definite choice is forced on it by how we make the measurement. Till then, there apparently is no "real" electron but only a superposition of quantum states for both wave and particle. This half-and-half phantom, state is illustrated by the paradox known as the Schroedinger cat experiment. Erwin Schroedinger, developer of the well-known QM wave equation, proposed the following gruesome thought experiment to illustrate the bizarre consequences of QM theory. The radioactive decay of a specific atom, which we cannot predict directly, is set to release a poison gas into a box containing a cat. We arrange for a 50/50 probability that this will happen. Is the cat dead, or is the cat alive? Or perhaps the cat is in a ghost-like, half alive and half dead, state till we collapse the wave function by looking in the box?

Part of the "shocking" nature of QM is that there are reputable people who take this question most seriously. No wonder Einstein refused to accept the reality of QM and fought against the standard interpretation all his life. Einstein, and those who supported him, proposed over the years a number of experiments which would yield impossible-seeming results if QM behaved as the theory suggests. Einstein and his collaborators argued that the standard QM model must be incomplete because it could not explain certain results. How is it possible, for example, for a measurement on one part of a system to have an effect on another part of the system when there was not enough time for a light-speed signal to travel between these? This so called "nonlocal behavior" and other results convinced Einstein that something fundamental was missing from our understanding. He felt that the randomness of the results was like "playing dice," and he would not accept it. Many of Einstein's suggestions, and a number of additional paradox creating ideas, have been tested in the last half-century. The results have always been in accord with classical QM. Nonlocality is real. The randomness analogized by Einstein's dice is real. No one can explain on a most basic scientific level why the universe behaves this way. Those of us, however, who believe in the Bible do have an answer. This essay shows at the outset that these strange results are in accord with Torah. Indeed, that Torah requires the universe to be this way. Let us take a more detailed look at some of the strange behavior of the QM universe.

The double-slit experiment

Richard Feynman says, in his lectures on *The Character of Physical Law*, that nothing illustrates the bizarre nature of QM as well and as simply as the double-slit experiment. The first thing to note about the strange nature of QM at its most fundamental level, is that electrons behave in exactly the same way in this experiment as light quanta. How is such a result possible? After all, light and electrons are obviously different in a number of fundamental ways. Electrons have mass, while light quanta have no mass. We cannot assign a size or structural designation for light, while we speak of a radius of 2.82×10^{-13} cm for the classical electron. Electrons can move at a wide variety of velocities. Light is constrained to the speed of electromagnetic radiation, commonly known as the speed of light. Nothing faster and nothing slower is permitted. Electrons experience time duration just as all other matter in the universe, including ourselves, are subjected to the flow of time. Light quanta, moving at the speed of light, do not experience time duration. These exist at a perpetual "now."

Light quanta and electrons are clearly different in many fundamental ways. So why is it that these yield the same shocking, bizarre, strange, unbelievable, spooky … – well you get the idea – results? And if this is not enough, consider that fundamental QM theory would have us believe that larger stuff, such as Schroedinger's cat, ourselves and even the whole universe, will behave the same way too. What does this mean? Science does not know. Torah suggests some answers, as discussed at the beginning of this essay. But Torah also does not really explain the fundamental results. Rather, Torah tells us why the world has to be structured in such a way, and why it is that there are things in this universe that we cannot know.

The basic structure of the double-slit experiment used to demonstrate the QM behavior of photons or electrons, is essentially the same as that used in 1801 by Young to demonstrate the wave nature of light. This time, however, we will perform the experiment several times under somewhat differing conditions. First we will use indivisible particles, then waves and then electrons or photons. We will see from the results that electrons and photons exhibit properties of particles and properties of waves, but are neither pure particle or waves. The critical part of the apparatus is a source of bullets or electrons or light quanta, a screen with two slit-like openings through which the particles or light can pass, and a catcher screen in the back to show what gets through the slits. Some systems also have a screen in front with a small opening to collimate the beam of light or particles. This is the process suggested by Feynman, and this is how I will describe it in this essay.

The apparatus consists of a screen with a small opening, behind which is another screen with two small openings close together. There is a third screen in the back, whose purpose is to catch whatever gets through the openings, or slits, in the middle screen. We begin by shooting particles through the opening in the first screen. These particles have some special properties. They are all identical, they are small, they cannot be broken up into smaller pieces, and they are emitted with the same velocity. Real bullets don't behave this way, but we can get close enough to easily perform this experiment. The source or gun of these bullets is not steady so the particles don't all land on one spot. Furthermore, some of the bullets are deflected off the edge of the first hole, so they move out of line even further. As a result, some of the bullets are stopped by the second screen, but some go through the openings of the second screen. These get caught or embedded in the third screen. We shoot a lot of these bullets, and we look at the third screen to see how many got through and where they landed. We see a pattern. There is a clump behind each of the holes in the second screen with smaller numbers of bullets embedded further out. We now close one of the holes to see what we get. We then open this hole and close the second hole and see what we get. We compare the three results: hole 1, hole 2 and both holes. We see that the result for both holes open is the sum of each hole by itself. Each hole is independent of the other one. They don't care whether there is or is not a second hole open. We get the same result for each hole, and we can add the results for each of the holes to find out what happens when both holes are open. It is a simple result, and it fits our experiential perception of reality.

We now perform the experiment with waves. Any waves will do. The simplest arrangement is to use water waves with obstacles in the water, using openings in the obstacles to make the slits. Waves, unlike bullets, can go around a corner. The phenomenon is called diffraction, and that is what Young saw for light. Waves also superimpose, or add. We get a bigger height where two crests come together, and we get a cancellation where a crest meets a trough. This phenomenon is called interference.

We have constructive interference when waves add in the same direction, and destructive interference when they subtract. With both slits open, the result is bands of high-level waves alternating with bands of almost nothing where the waves cancel each other. As expected, the interference bands disappear when we close one of the holes. We add the results for each hole on its own and compare them to the result with both holes open. The numbers do not fit. The result with both holes open is not equal to the sum of each of the holes on their own, as we had for particles. The result can be calculated, but it is a more complicated equation. The result depends on the squares of the intensities, so a trough no longer directly cancels a crest. The behavior is not as simple or intuitive as for the bullets. But a bit of work shows where everything comes from, and why it works that way. There is nothing hidden. There is no mystery here.

Now we perform the same experiment using either light photons or electrons. Both holes are open. We shoot one photon at a time, very slowly. We notice a spot on the third screen, and we surmise that the photon came through one of the slits in the second screen. We do this many times. Finally a pattern emerges on the third screen. It is an interference pattern indicative of a wave. But you need to have the wave go through both slits simultaneously to cause interference. Does this mean that a photon can break into two parts and go through both slits? Well, perhaps photons are not indivisible as we had thought. So let us try this experiment with electrons. We do it and we get the same result as for the photons. The electron, which has mass and seems to be a particle in all other ways, is behaving like a wave. Not only that, but a single electron apparently transits through both slits simultaneously. How is that possible? We don't know. Are electrons then really waves? No, because we get the result for particles when looking at one hole at a time. It is only when both holes are open that we get the result for waves.

The result for the electron or photon is not just strange, but very strange. Consider the following modification to the experiment. We will observe the probability of arrival, or number of electron hits per unit time, at a specific spot on the third screen, the collector screen. We get a certain value with both holes open. With one hole closed we get a larger value. Switching closed and open holes yields a smaller value. Or if we pick just the right spot, we can find that having both holes open will yield a bigger result than either hole by itself or the sum of each hole by itself. This says that somehow the electron or photon going through one of the holes, knows when the other hole is open or closed. How does it know that? Let us look further into this mystery.

A QM view of the double-slit experiment

QM represents the electron by a probability wave function. The electron, or rather its quantum wave function, takes all possible paths as an electron moves from point A to point B. We discover where the electron lands by summing the squares of these waves, using appropriate phase relationships for addition or cancellation. The electron has two possible paths in the double-slit experiment, hence there are two quantum wave functions to combine. Viewed this way, we could say that the electron does indeed split into two parts to pass through both slits. Indeed, we could conceive of an arrangement involving more than the two paths which the electron takes. The multiple paths are there because we do not know which way the electron actually went. Are there still multiple paths if we do know, by measurement, which way the electron goes? The answer is no. We do not have to close one slit to eliminate the interference pattern. All we need is to find out which path the electron took, and the other paths no longer exist for the electron.

The observer, that is us, is part of the system in QM theory. Einstein did not care for this and proposed various tests to show that the underlying reality of the universe is more conventional. But all tests performed so far verify that the universe does behave in a classical QM way. The results of these experiments show that QM behavior is deeply embedded in the fabric of the universe. Light quanta are not pulsed waves, as a simple-minded analogy earlier in this essay tried to suggest. Light quanta (and electrons, and possibly everything else) is neither wave nor particle nor anything that we have had direct experience with. We are not precluded from making certain measurements because our methods are too crude. And we are not prevented from calculating, or predicting, certain results because our theories are missing information – an idea known as the hidden variable hypothesis. It is not possible to conceive of an instrument that can make the measurement, and there are no hidden variables (at least on a local basis). The universe is just not deterministic for us.

The multiple paths of the electron or photon, that somehow know about each other, are called entangled quantum states. Numerous high-precision, cleverly elegant experiments have been performed involving versions of entangled quantum states. As noted already, these have all shown agreement with standard QM theory. Suppose we shine a light behind one of the two slits. We detect the collision of the electron with a photon when the electron goes through that slit. Being able to determine the path of the electron destroys the interference. A more advanced version of the above experiment was introduced by John Wheeler. The collecting screen can be rapidly retracted to uncover two telescopes, each focused on one of the slits. These are used to determine through which slit the photon passes. We get the interference pattern of a wave when the screen is in place, but the individual particle pattern when we determine the particle path. This happens even when the choice to move the screen and uncover the telescopes is made safely after the photon has passed by the location of the middle partition with the slits. The photon somehow retroactively makes an adjustment to behave as a wave and go through both slits, or act as a particle by going through only one slit, depending on whether the screen is up or down. Remember that the decision to move the screen was made only after the photon had passed through the slits. Even more complicated versions of this idea have been reported in the literature. The result is invariably the same. The past is adjusted so as to get the needed result in the future. How does this communication to the past work? We don't know.

More QM tricks

One trick of QM behavior involves communication between separate, but entangled parts of a quantum system, to force results in one part of the system to correlate with measurements of another part of the system. The previous section shows an example of communication to the past to explain measured results in the future. A more common example of such communication involves action at a distance. There are a number of experiments which show communication across distances that requires faster-than-light speed signaling. A simple version involves the measurement of some parameter at a specific location such as electron spin direction. Distant parts of the system will also assume this behavior, even though the result is measured virtually simultaneously with the action taken. In theory, one could do something on earth to affect the Andromeda galaxy without waiting 2 million years for a light-speed signal to get there.

Einstein was very unhappy with such nonlocal behavior. The argument was that something was missing from the theory. A QM variation involving "local hidden variables" was proposed as a solution to the impossible action-at-a-distance results. Recently this idea was tested, involving calculations of results of what is known as Bell's theorem. We do not need to know the details, except for the result. QM does violate locality. There are no hidden local variables. Nothing has been proven yet about nonlocal hidden variables. So possibly something can still be found to make this behavior a bit more palatable. But there is no way out of the basic dilemma. Photons can communicate with each other, tell each other their states, and indicate how to behave, at faster-than-light speed. But photons travel precisely at light speed. So how does this work? We don't know.

We do not know how it works but we know from recent experiments that while the behavior of entangled quantum states is indeed "the way predicted by quantum mechanics. However, the experiment also showed that there is no faster-than-light signal … [because rather than communicating, the different locations simply] share the same existence, the same destiny, if you like." It sounds impossible , but that is what recent experiments show as reported in the February, 2001, issue of the *IEEE SPECTRUM* in an article about "The Topsy Turvy World of Quantum Computing." This article provides a progress report on the latest technological adaptations of QM science. We are told that "quantum computers are no longer seen as curiosities but as the powerful future of the computer industry." We are told that IBM, Oxford University, Hewlett-Packard, Lucent Technologies, AT&T, Microsoft, U.S. National Institute of Standards and Technology, MIT, Caltech, University of New South Wales (Australia), Delft University of Technology in the Netherlands, University of Innsbruck (Austria), and other organizations, are racing to be the first to build a practical computer based on the "ghostly dual existence" of particle states in QM, and "by exploiting the phenomenon of quantum interference." Indeed, a new start-up company, MagiQ Technologies, aims to soon make a profit from the exploitation of QM phenomena. QM behavior is not conjecture or theory. QM behavior, bizarre as it seems, is reality.

The Bohr description of the atom includes the requirement of specific electron orbit locations. Electrons can change location by absorption or release of some energy. But an in-between position is forbidden. So what path does an electron take when changing position? The answer is that no path is involved. The electron disappears from one place, and an identical electron appears somewhere else. This is old stuff. Any college text in physics will have this information. Recently, ideas have emerged to show that a quantum state can be shifted, transported, teleported, or whatever word one chooses to use, from one place to another. The quantum state, as has been discussed before, represents the item or object being described. The physical entity appears when the quantum wave is collapsed. Teleporting a quantum state is in many respects equivalent to teleporting a physical object. Just like Bohr's electron – now you see it, now you don't. This process is not exactly in the technology exploitation stage. But it is not as impossible as it sounds. The lead article in the April, 2000, issue of the *Scientific American* magazine is titled "Quantum Teleportation, The Future of Travel? Or of Computing?" This is not just a common popularization. The fact that a major journal, the *Physical Review of Letters* (March 29, 1993), would publish a serious paper with the word "teleportation" in the title says something about where we are in the ongoing process of coming to grips with the bizarre nature of QM. QM teleportation experiments are no longer unusual. And those wishing to learn more about quantum teleportation, will find a report on the latest results on web site: http://info.uibk.ac.at/c/c7/c704/qo/photon/_teleport/index.html.

I mention, elsewhere in this book, the use in electronics of QM-based devices, including a very fast switch called a tunnel diode. In this device electrons are said to "tunnel" through a forbidden zone called a barrier. A calculation by Eugene Wigner and Leonard Eisenbud showed that the transit time to cross a barrier can sometimes be a constant, regardless of the length of the barrier. This means that the transit time can violate the speed-of-light limit of Relativity theory if the barrier is wide enough. Recent experiments appear to confirm this calculation. One experiment, by Professor Nimtz at the University of Cologne, reports the transmission of Mozart's 40th symphony at several times the speed of light. There is much discussion about the meaning of these results and whether the speed-of-light limit was indeed violated. The interesting point for this writer is that most people find Relativity to be strange enough. Now suddenly, that seems ordinary, and scientists agonize over whether QM somehow violates Relativity rules by exceeding the speed of light through tunneling and action at a distance.

Beyond QM tricks

The reader should by now be convinced that QM is very strange. But there is more that is stranger still. Nothing is not nothing. A vacuum is not nothing. It is something that exists. Furthermore, on a quantum level, there is stuff that exists in a vacuum. Quantum theory does not deal with certainty, but with probabilities. Nothing exists with absolute certainty. It's just that the reality we perceive has a very, very high probability associated with it. Stuff, energy and/or particles can pop out of nowhere according to QM theory. We don't see it because the probability is negligibly small. But the probability goes up as the amount of energy, or mass equivalent, and time of existence gets small. Space and time is huge, compared to these tiny things. QM vacuum fluctuations continually bring new stuff into existence, which lasts for a bit and then disappears. The vacuum is a busy place, with churning chaotic appearance and disappearance of energy spurts. This is not just a theory. Experiments using the Casimir Effect, predicted by Dutch physicist, Hendrick Casimir in 1948, measure the force generated by so called "virtual particles" that pervade "empty" space. These "virtual particles" are observed, measured and catalogued as part of the collision process in high-energy physics colliders.

What if the temperature and pressure were not just high, as produced in current laboratory collision experiments, but super high, as it was in the early universe? Then all sorts of stuff is possible. And theory says that this is how it happened. The inflationary scenario of the big bang theory explains many observational factors, including the uniformity of the cosmic background radiation. This theory also suggests that all of the energy and mass of the universe is due to a false vacuum fluctuation. The "false vacuum" was there before the real vacuum, the physical vacuum which we now have, came into being. All the stuff came out of nothing. It is a nothing that is beyond nothing, and we have no common language to express this. The difficulty of expressing such ideas in common language is not new. Thus, Nachmanides explains on page 23 of his commentary on Genesis that, "He created all things from absolute non-existence ... He brought forth [the universe] from total and absolute nothing ...." But this is not a complete explanation because "We have no expression in the sacred language ..." to directly explain this.

We can go even further. Attempts to formulate a quantum gravity theory show that space-time itself can be a quantum fluctuation. But what is it that fluctuated, if there was nothing in existence, not even space or time? We have the false vacuum and farther back beyond this there is a so called "multiverse" on which our universe is just one of many bubbles. The inflationary theory, originally suggested by Alexei Starobinski and developed into its current form by Alan Guth, says that we are but one of an infinity of possible universes. This is an outgrowth of strong anthropic principle reasoning and classical QM theory. Martin Rees makes it clear, in *Just Six Numbers*, that he finds this idea most attractive. Thus, we are told on page 150, that "I think the multiverse genuinely lies within the province of science, even though it is plainly still no more than a tentative hypothesis."

Having one serious contender for a "multiple universes," a multiverse, idea is bizarre enough. But there is more. We have a totally different, equally serious science-based proposal that we are just part of an infinity of universes. And yes, this suggestion is also based on the peculiar notions of QM theory. Is the idea that not only space and matter, but time itself was originally created an issue for Torah? Of course not. Thus, Maimonides states in chapter 30 of Part Two in *The Guide for the Perplexed* that "The universe has not been created out of an element that preceded it in time, since time itself formed part of creation … ."

Orthodox QM treats objects as waves which collapse into an object when the wave interacts with an observation or measurement. A mind acquiring information about the wave-like object collapses the wave into an object. This explains the results of the two-slit experiment. An electron can go through either of the two openings, and this introduces two QM wave representations for the electron. The interference that we measure is between these two waves. But the waves collapse into a particle, which we call the classical electron, once we determine through which slit the electron actually passed. The classical, also known as the Copenhagen, understanding of QM involves the mind of the observer. And that bothers some people. A different interpretation is provided by John Wheeler and Hugh Everett. This is known as the Everett-Wheeler "many worlds" interpretation of QM. This interpretation posits that there is no collapsing of wave functions involved. The electron takes every path that it can take. In the two-slit experiment, this means that the electron goes through both slits.

How is that possible? Simple, the universe splits into two different universes. The electron goes through one slit in one universe and through the other one in the other universe. The universe is continually splitting into multiple universes every time there is some decision point somewhere. Considering the size of the universe and the number of decision opportunities, this makes for a lot of splitting. In effect, we have virtually an infinite number of universes corresponding to all variations of all decision outcomes ever made. Sound like science fiction? No, this is real science. QM results made via the many-worlds interpretation fit reality just as well as the classical procedures.

Currently we have two multiple universes theories. Both are derived from QM processes, but the theories lead to different results. The false vacuum with bubbles multiverse postulates separately isolated bubble universes. We would need access to the underlying mother multiverse to move from one universe to another. The many-worlds system, however, has a continuous splitting off of universes via each decision process. There is a decision-based causal connection between universes, and there is no higher order multiverse involved. Which theory is correct? Is either theory correct? Who knows? Possibly science will come up with some means to test these ideas, but nobody knows how to do it at this time. That does not mean that there is no interest in a multiple universe structure. Such an arrangement solves some of the nasty problems associated with the probability that the universe is as it is. How could something with such a low probability come about? Easy, the probability is not low. There are an infinite number of universes. A probability of one in a googol, or a googolplex or any other number is not a problem when compared to an infinity of chances. Multiple universes is not a problem for Torah, should such a theory one day find corroboration. Commenting on Talmud *Avoda Zorah* 3b, which mentions that there are multiple worlds, Rabbi Yehuda ben Barzilai of Barcelona explains that these are not planet-type worlds, but rather each world is a completely separate universe.

Finally a comment about reference sources for the reader. The July 20, 1998, issue of *U.S. News* magazine has an extensive discussion on the topic "Is Ours The Only Universe?" In an interesting coincidence, the story tries to give an analogy to the beginning size of our universe. The beginning is called a "cosmic mustard grain." It's a catchy analogy. But the idea is not new. The Torah commentary compared the beginning universe to a mustard seed many hundreds of years ago. For those interested in a deeper look, a good source for the more out of the way peculiarities of QM, and other physics as well, is available on the internet at the web site of University of Washington physicist, John G. Cramer. The URL is www.npl.washington.edu/av.

Chapter 9

What Are We?

Man in Science and Torah

Man is the pinnacle of creation in Torah. The universe has no purpose except as a place for us to exercise our free will in the service of Hashem. Thus has it been understood since the Torah was revealed at Sinai. Man is not a result of the existence of the world. Rather, man is the cause and the purpose for the existence of the world. But science, apparently, does not see things this way. To science we are just a result, perhaps even a random and accidental result. There is no plan. There never was one. The universe exists by accident. Life came about by accident. Life developed one way, and it could just as easily have gone some other way. What is the difference? The accidental, almost incidental, mutations of evolution led to human beings. So what, and who cares? We care, of course. But we do not count. We did not make this happen. A stray asteroid hit the earth and wiped out the dinosaurs, apparently. This opened up an evolutionary niche for the development of creatures like us. We would not be around if it were not for that asteroid. Instead the dinosaurs would still be here. And it is all the same either way. Man or dinosaur, it makes no difference.

There appears to be a major difference in perception between science and Torah here. It looks as if we have a disagreement without any possibility of reconciliation. Indeed, this was the case in prior times. But it is no longer so. Today, the conflict is only on a superficial, on a surface, level. A deeper look at the latest science actually shows a remarkable degree of commonality between science and Torah. The new science shows the universe to be chaotic and non-deterministic as compared to the orderly structure of Newtonian mechanics. Yet paradoxically, the less we can predict about the universe, the more important the position of mankind has become. Two current scientific modes of inquiry, anthropic thinking and quantum mechanics (QM), have significantly enhanced the status of mankind in the hierarchy of the universe. Indeed, when taken to the ultimate limit, the universe apparently cannot exist without us. Let's take a closer look at what this is about.

The anthropic principle

Most people are not familiar with the anthropic principle, of which there are a number of versions as is mentioned in an earlier chapter: the weak, the strong, the participatory, and the final. Only the better known, weak and strong, versions are discussed in this essay. The reader will find a great deal more in paragraphs and chapters in various books on science, and in a book devoted solely to this subject, *The Anthropic Cosmological Principle*, by John Barrow and Frank Tipler.

A scientific principle is not in the same category as a law, or theory or hypothesis that tries to explain the workings of nature. Rather, it is a mode of investigation and analysis that leads to laws and theories. One of the more frequently invoked scientific principles is known as Occam's razor. This principle states that among several explanations of facts, the simplest explanation is the most likely to be the correct one. Another example is the anthropic principle. A scientific principle that is anthropic must have humanity as a prime partner, given that "anthrop" is another word for human being.

A basic statement of the anthropic principle might go as follows: the reason that the universe is the way it is, is that we are around to wonder about it. This sounds a bit mysterious. Why such an idea? The reason is that we know the universe is very delicately balanced in many of its laws, forces and structures. The tiniest change in any direction, with probabilities running into the googol category, and we would not exist. The question then arises – how is such a thing possible? How come the universe is this way? The answer is that this is the only universe that we can observe or wonder about. We cannot possibly observe and wonder about the existence of a different sort of universe, because we could not exist in a different universe. The anthropic principle is simply a statement of fact, when taken this way. It does not explain anything or lead to any new insights. We simply note that the universe we observe has the properties that it has, because this is the only kind of universe where we can exist to observe anything. But this statement of fact is only the beginning. There is more to the anthropic principle than that.

The anthropic principle appears to be neutral when taken as a statement of fact. But a neutral statement has no scientific validity. Hence, there are those who consider it useless, if not actually ridiculous, to invoke this principle. Consider the following analogies. You ask someone why he is walking. The response is, because he is not standing still. You ask why a glass is empty, and you are told that this is because the glass is not filled. You ask why we observe this version of a universe, and you are told that the reason is that it is not some other version of a universe. The universe is the way it is, because that is the way it is. Of what use is that? Heinz Pagels (*Perfect Symmetry*) says that the anthropic principle "has explained nothing," and has had "a negative influence" on science. A more sophisticated variation does lead to new modes of thinking, and new hypotheses about the universe. But these can get very complicated, such as the "many universes" idea, for example. Hence, there are those who do not care for this variation either, because the results violate Occam's razor principle of simplicity.

Not all scientists support anthropic analysis, and in fact, there is much criticism of anthropic thinking. Does this criticism also hurt the cause for Torah thinking? Certainly not. Remember what this principle is about. The anthropic principle is an explanation in response to the impossible probability that we have this universe with this life in it. Torah is not damaged if science cannot explain logically why or how this universe can exist. The Torah tells us why, and the believer does not need scientific validation. In response, critics of the Bible will sometimes ask where G-d came from. The answer is that we do not know. So what? Science also does not know. An alternative version is that science "knows," but the answer is not exactly simple. The strong anthropic principle postulates a multiplicity of universes growing out of a higher-order mother universe. Where did this mother universe come from? Science will agree that we do not know. If science cannot explain this universe as "natural," then what is left if not the Bible? Torah is not damaged either way.

There are two primary variations of anthropic principle, strong and weak. The weak version provides insights to the structure and laws of the universe based on our knowledge that we exist. We know that the universe must fit this existence. The strong version goes so far as to call for a universe that includes conscious beings in it. In other words, we not just happen to exist, but we must exist. The argument for this idea was first proposed by astrophysicist Brandon Carter in 1974. This idea is supported by the philosophy of logical positivism which argues that it is meaningless to discuss that which can never be observed. What does it mean to have a universe that we cannot observe in any way? Is that any different from a work of fiction? Torah is in agreement about the need for intelligence in the world. But it does not require human observers. Alternate universes that do not include intelligent beings is not contrary to Torah, given that Torah explains all as a creation by the ultimate Observer. Is anthropic thinking, then, just a fictional game of make-believe as argued by the critics? Not so, say the supporters. It is true that this principle does not explain why this universe exists or why we exist. But many important aspects of our universe become obvious, once we accept that the universe has to fit our existence.

Why does the universe have three dimensions, and not two or four? Gravitational attraction would cause unstable orbits for a planet around a sun in more than three dimensions, and living beings need a stable planetary environment. Two-dimensional creatures cannot have internal openings for nourishment or waste elimination because these would cut the creature into pieces. The thermodynamic arrow of time, involving an increase in disorder, is needed so that we can take in food which is structured and ordered and convert this into movement, heat and disorder. Living organisms convert order to disorder, and this gives a time direction. For us to exist, it is necessary for the universe to be old enough to form stars; for the stars to go nova at the end of life and spew out ashes including carbon; for new stars to form that include carbon; for basic life to form; for life to evolve from simple to complex, and so forth. Reasoning of this sort shows that the universe has to be at least 10 to 15 billion years old, because that is the minimum time needed for creatures like us to come into being. On the other side, we know that the universe is probably not more than 20 or at most 30 billion years old. This is more than enough time for intelligent life to evolve according to anthropic reasoning. We are intelligent life, and we exist. Anthropic thinking indicates that we exist right around the lowest time frame when such life is possible. This may explain why ET has not come visiting. The anthropic explanation is that we are the first and highest form of life in the universe. These are some of the relatively simple conclusions flowing from the weak version of the principle.

The strong anthropic principle is more direct in promoting the importance of intelligent life. It requires that intelligent life must exist. This is a problem, because statistical calculations show that the probability of getting our kind of universe, and our kind of life, is virtually impossible on a random basis. This leads to the suggestion that there are an infinite number of different universes, one of which happens to be ours. A probability of even one in a googolplex is sufficient to explain the existence of this universe under such conditions. The "many universes" idea is attractive for a number of reasons. One is that the inflationary big bang hypothesis suggests multiple universes based on quantum fluctuations in the false vacuum that preceded the normal, or physical vacuum, of empty space. Inflation does not require this, but it does suggest it. Inflation at the initial stages of the universe is now well established because the idea explains, and fits well with experimental measurements of, the properties of the cosmic background radiation. Those who favor the strong anthropic principle use inflation theory in support of their position. The argument is that this principle rather than having "explained nothing" as is suggested by critics, has in fact explained one of the results of inflationary theory.

The most important support for the strong version of the principle, however, has nothing to do with science. Rather it is a philosophical preference for a system that explains our existence without invoking a Creator. Stephen Hawking is not a supporter of the Bible. He makes it clear in his book, *A Brief History Of Time*, that a more advanced scientific understanding will eventually explain everything. At this time, however, we have only two logical choices in explaining the existence of the universe and ourselves. "One can take this either as evidence of a Divine purpose in Creation … or as support for the strong anthropic principle." Either the universe is an "act of G-d who intended to create beings like us." Or, we accept the strong anthropic principle, because "the strong anthropic principle would claim that this whole vast construction exists simply for our sake." Torah is happy with either explanation. The anthropic principle is not a threat, because multiple universes are perfectly acceptable to Torah. The only thing required by Torah is that "this whole vast construction exists simply for our sake."

Quantum mechanics

QM started as a description of the small, of the very small, such as an electron. The world of the small is a world of uncertainties and probabilities. Both the uncertainty and probability are associated with an observer making a measurement. We start with the uncertainty principle, formulated by Werner Heisenberg in 1926. This principle states that certain pairs of properties of a particle, such as position and velocity (actually momentum), are interdependent. The more accurately you determine one, the less you can know the other. The product of the uncertainties cannot be less than a certain value connected to Plank's constant. This is not a consequence of inadequate instrumentation, but rather a fundamental property of the universe. Improved techniques, better test instruments, and increases of the amount of time spent taking measurements will not help. Furthermore, we cannot cheat by taking a measurement twice, one for position and a second measurement for velocity. When you make the second measurement, velocity in this example, you not only cannot determine the position, but you actually affect the position. Hence, the previous measurement of position is no longer applicable. There simply is no way out of this dilemma.

Indeed, the particle, say an electron, may not even have a unique position or momentum till it is measured. This is the basis of the standard model of what is known as quantum mechanics as distinguished from ordinary, or Newtonian, mechanics. The standard QM model, also known as the Copenhagen interpretation, was developed by a number of scientists, with the leadership role attributed to Niels Bohr. In this model, the position or velocity of the electron are no longer to be considered separately but described by a combined quantum state. Computations involving mathematical procedures or relationships, such as Schroedinger's equation, will not yield a definite result but only a probability. The position of an electron is described in a smeared out, distributed, state by a probability distribution. The question arises whether this is just a mathematical artifice for purposes of computation, or whether it describes the real world? Einstein objected strenuously that this description is not how the world actually works, leading to his oft-repeated quote about a universe governed by rules of a dice game. But the universe does appear to follow the probability rules of a dice game. All experimental tests to date show results that agree with the standard QM model.

The electron has no real existence in QM formalism at least not the sense we normally mean by the word "real." The electron is spread out in a wave, or multiple waves involving all paths that a moving electron may take. There is no physical electron. All we have is probability information telling us that if we look we are likely to find an electron in certain locations. The wave representation collapses when we look, and we find the electron in one of these locations. But this only happens when we look. It takes an observer to make a physical matter electron. Till then all there is, is a wave. Not only that, but there are many waves as the electron moves. We have to find the superposition, or sum, of these waves to find the real electron. And the summation over time is not done with the mathematics of the real time-domain that we experience, but in the domain of the mathematics of imaginary numbers involving the square root of minus one. We get the correct answer only when operating in the realm of imaginary time. Is it possible that real time is what we experience, but imaginary time is how the universe really behaves? It is possible, says Stephen Hawking in his *A Brief History of Time*.

The QM behavior of energy and matter on the small scale is bizarre enough. But now comes the really crazy stuff. Standard QM understanding says that this behavior applies to everything, whether small or large. It applies to an electron, it applies to a chair, it applies to humans, it applies to planets, it even applies to the whole universe. What does it mean to say that a chair, a person, a planet, and the universe are all just probability waves without solid existence? Only the act of observation, only the act of measurement causes a real existence to come about. Does that mean that a chair does not exist as a solid chair in one location unless someone looks at it? Yes, that is a possible conclusion. Does it mean that the universe was only a wave function till someone looked at it? Yes, that is what it can mean. Who then caused the wave function of the universe to collapse? The Torah has an answer, but science does not. No wonder the many-worlds QM interpretation, which does not require the collapsing of wave functions, is so attractive to some scientists. We can go even further in this direction, if the above is not sufficiently strange.

John von Neumann has analyzed what happens to the QM uncertainty when a measurement is made. He has shown that the ambiguity, or uncertainty, or spread out quality, or whatever we choose to call it, is removed from the item being measured and transferred to the measuring apparatus. So why do we perceive a non-ambiguous measuring apparatus? Because the process of transference continues as we observe the measuring apparatus. The ambiguity is removed from the measuring instrument and transferred to ourselves. And what happens then? Says leading QM theorist, Eugene Wigner, the uncertainty disappears as it interacts with a conscious mind. Apparently, the conscious mind removes uncertainty and creates reality. It is a strong affirmation of the strong anthropic principle. This idea says that the universe exists because we are here to observe it. But wait. We cannot observe the whole universe. Does that mean that those parts of the universe we do not observe do not exist? Torah has no problem with that. The whole universe and every particle and every quantum fluctuation is observed and directed to behave as it does. Which is easier to accept on faith? The Torah view, or that we are the creators of the universe, and our own creators as well?

The Torah view

A prayer that we say daily begins with the words, "What are we?" The prayer then proceeds to list the deficiencies and frailties of man, culminating with the statement that "The pre-eminence of man over beast is nonexistent ... except for the pure soul...." People have a body. It is a uniquely structured body. So what, and who cares? Rabbits also have a uniquely structured body. The tiger has a uniquely structured body. The horse, and elephant, and penguin and all other creatures have specific bodies. Is the shape of the body what gives humans their special status? No, we are all equally "living creatures," *Nefesh chayah* per Genesis 1:24. Our uniqueness comes from the human soul, and not from the body, *Nishmat Chayim*, "the soul of life," per Genesis 2:7. The Torah agrees with Wigner about the primacy of humanity, though support for Torah is not what Wigner had in mind when he suggested his hypothesis about the importance of human consciousness.

The Torah, especially the first sections of Genesis, is written in a terse and cryptic style. The basic statement provides only the bare minimum that we need to know.

Over and over, the commentators explain why. The Torah is written in the language of man. But some things do not lend themselves to such language because they are beyond our capacity to understand. Nevertheless, there are deeper levels beyond the basic or surface meaning so that at all times and in all generations people would be able to understand Torah at their level. Armed with 20th century science, we are able to discern meaning which people millennia, or even centuries ago, could not fathom. The Torah is complete, and nothing is missing that human beings can handle on a practical or potential basis. But, as with a growing child, to whom some advanced and sophisticated matters cannot be taught till the child grows older and wiser, so is it also with us. The deepest secrets of Torah remain hidden, and the material comes out only when people are ready to deal with it. Except for a very few exceptional people to whom the book is fully open. These have access to the esoteric and secret knowledge which is not publicly taught or discussed. Respecting matters of creation, we are told that a teacher may teach this only to a maximum of two students and only if these are exceptionally qualified. The commentaries that ordinary mortals, like this writer, have access to are only a pale reflection of the deepest teachings.

Today, owing to the preparation of the general populace by science, we start to see glimpses of Torah teachings that go even deeper than the factors of creation. These are commonly known as Kaballah. Here the material goes beyond the confines of what science can deal with. At one time, that last sentence would have made no sense. But now we know from QM that science is limited in what can be known of the universe, even in principle. Yet Torah is not science. Torah transcends science, and it can probe deeper. But even Torah has its limits. We are told specifically that we can never know the ultimate reality. Moses, the greatest prophet, realized that he did not have access to all, and he asked for more. The response was that, "... [Y]ou will see My back, but My face may not be seen" (Exodus 33:23). Einstein, from the perspective of science, like Moses from his perspective of prophecy, realized that we do not have access to the ultimate reality. Basic QM works superbly in manipulating numbers to get results. But a probabilistic universe denies us access to the basic principles explaining "why." Einstein refused to accept this. But all tests show that this is how the universe really works. We can see the back of the structure, but we cannot see it directly. Looking at the deepest levels, we find that Torah and science are in full agreement.

Humanity is the pinnacle and purpose of creation. The whole universe exists solely for our sake. As early 18th century commentator Rabbi Luzzatto explains in his *The Ways of G-d*, the "essential creature is man. All other created things ... only exist for his sake, to complete his environment ...." How man fits into the universe can be understood from Genesis, with a little help from the commentaries. As noted, only the most superficial and childishly simple understanding will be obtained from reading the opening parts of Genesis, without commentary. Fortunately, we need not delve into the deeper and more difficult to understand commentaries to fit the creation of mankind into a scientifically understandable sequence. There are numerous commentaries on this matter, all pointing essentially in the same direction. What follows will include some quotations from just one of these, 11th century commentator Rashi, whose writing is considered a model of simplicity and lucidity.

We have several statements in Genesis involving the creation of man, (this term refers to man as in all human beings, man as in the male person, and finally the female person.) We need to examine the following statements in Genesis: "Let the earth bring forth living creatures [*nefesh chayah*], each according to its kind" (1:24). "Let us make [*asah*] man in Our image [*tzelem*], after Our likeness [*demut*]" (1:26). "So G-d created [*barah*] man in His image, in the image of G-d He created him [*barah*]; male and female He created [*barah*] them" (1:27). "G-d formed [*yatzar*] the man of dust from the ground, and He blew into his nostrils the soul of life [*neshamah*]" (2:7). Starting with verse 2:18, we are told that "it is not good for man to be alone." Later the animals are brought in, and the man names them all. "[B]ut as for man, he did not find a helper against him [*ezer k'negdo*]" (2:20). The next several verses describe how the man (that is, the generic human being) was placed into a deep sleep, and flesh was removed from one of his sides (some translate the word as rib), and from this was made or built (*v'yiben*) a female person. When the man awakened from the anesthesia, he was no longer a generic human being, but rather a male human being (man), and his counterpart, his helper against him (*ezer k'negdo*), was a female human being (woman). The original human being (man) was fashioned from the earth, but male and female, man and woman, came from the same living source. They were "bone of my bones and flesh of my flesh." And when they join together, they again "become one flesh" (2:23, 24). That is all we have. But it is quite enough for the commentators to explain everything.

The above narrative can be confusing if we take it as a sequential description of events. That is not the case. Rather the same happenings are described several times in several ways. Referring to the creation of the man and woman in 1:27, Rashi notes that "[B]oth were created on the sixth day, and it does not explain to you how their creation took place, and it is explained to you in another place." The last statement helps explain the first one; the first statement clarifies the middle; the middle clarifies the end and beginning, and so forth. We need to deal with these matters as a unified whole. Furthermore, the narrative does not lend itself to straight translation. Variations in spelling, choice of synonyms, sentence structure, and so forth provide multiple simultaneous meanings to what is described. In addition, we need to place the creation of man within the context of all creation. Only by taking into account all of the above can we make some sense of what is described here.

There are three basic categories of creation, aside from the inanimate universe. Angels are perfect spiritual beings. They perfectly fulfill the will of their Creator, but as a consequence they have no free will to act on their own. Ordinary animals, the *nefesh chayah* of verse 1:24, consist of a living body without an intellect. They too have no free will, being subjected to the urges of the body and without any understanding of a higher purpose. A human-shaped living being also came into existence. Yet this was not a human being, but merely a human body shape possessing life. It was only with the addition of the human soul (*neshamah*) that people acquired the special characteristic of being in the image (*tzelem*) of the Creator. This is a being with dual characteristics, angelic and animal. This fusion establishes something new, a being that possesses the intellect to understand the world and recognize that there is a Creator, and the free will to choose and act without the body-based compulsion of the animals or the Heavenly-based compulsion of the angels.

Differing commentators combine the above elements in somewhat different ways. I have chosen to follow a path that fits the purpose of this essay, which is to compare science with Torah. And I am providing only the minimum material necessary to permit this comparison. Much more is available respecting these verses. The interested reader can examine various commentaries. These will provide much more material, but with respect to the matters at hand, the differences are only in minor detail. All are in agreement as to the implications of the final result. The explanations that follow are offered in a particular sequence. This is because writing is based on a sequence of letters, words, sentences, and ideas. It is an inherent limitation of the medium. Please keep in mind that all is interconnected. The first statement is not more important than the second statement. It is just that something has to come first.

The first statement about people is, "Let us make man." This is in plural form. Who are the ones who comprise the "us?" Was help enlisted in making man? The commentaries assure us that the answer is no. There is only one Creator. Thus, Rashi explains that the next verse says, "[A]nd He created man, and it (scripture) did not write, and they created." The actual creation is described in the singular and not in the plural. Then why the use of "us?" There are several responses, all of which can be combined into one general response. One reason is that the angels were informed that a new being was about to be created who would include a spiritual dimension, which was the realm of the angels. Another answer is that the Torah, which is the plan of creation and which predates creation, was consulted as to the proper place of man in the universe. A final explanation is that man himself was addressed about his/her future. The whole book, as is evident from the title, and especially chapter 16, of *Let Us Make Man* by psychiatrist Rabbi Dr. Abraham J. Twerski, deals with this matter. Suffice it to say that by virtue of his free will, and the gift of being in the image (*tzelem*) of the Creator with creative abilities of his own, man is invited to be a partner in creation. What we do with ourselves, and what we do with the universe, of which we are caretakers, will affect the final configuration of creation.

Man was formed (*yatzar*) from previously existing material (dust), "and He blew into his nostrils the soul of life [*neshamah*]." "Blew" is an anthropomorphic expression. The Torah is written in the language of man so that we may learn from it. As in the case with a person who blows breath from his body into someone else's body, so here some spiritual essence of the Creator was transferred to the created man. This is the human soul (*neshamah*) in contrast to the ordinary animal life (*nefesh*). The human body was formed (*yatzar*), but the human being with a human soul (*neshamah*) was created (*barah*). Furthermore, this applies to both the man and the woman. Thus, "[M]ale and female He created (*barah*) them." Some would associate the creation of *barah* with the human soul, while others say that the soul was made (*asah*) and the new creation is in the combination of non-material soul with material body. In either case, the human being is a creation of the highest order. We have the same *barah* as in the initial start of the universe, where we are told that, "In the beginning created (*barah*) ...." Human beings have been granted creative powers. We can shape inanimate matter and manipulate the genetic material of life forms. One day, we may even learn how to create totally new life, as in *nefesh chayah*. But the deeper realm of *barah* is beyond our ability.

Man was granted the gift of being in the image (*tzelem*) and likeness (*demut*) of the Creator through the infusion of a spiritual substance, the nishamah. This "image" and "likeness" is not physical, but only spiritual and mental. Rashi says that this "refers to our power to understand and to discern." We can manipulate matter and energy, we can discover the laws for the "natural" workings of the universe, we can contemplate and understand our position and purpose in creation, we can conceive that there is a Creator and understand some of His essence, and above all, we have free will to choose and to act. Animals cannot choose. They must follow what their animal instinct dictates. But we have two inclinations to choose from. We can choose what is proper (*yetzer tov*) in conformity with the highest standards of our spiritual nature, or we can choose to act improperly (*yetzer rah*) by giving in to the baser instincts of our animal side. This is alluded to by the variant spellings of the *yatzar* as animal and man are "formed." The animal version is spelled with one letter *yud*, whereas for man we have two *yuds*. Moreover, the placement of two *yuds* in sequence names Hashem. The difference in inherent nature of man and animal is a deliberate part of creation. Man can act like an animal, and man can act like an angel. But man is neither. We are a unique being, a partner in creation and its highest expression.

There is clearly a time gap between the creation of the single individual man, and the later separation into male and female, man and woman. Various events occurred during this time, such as the naming of the animals and the unsuccessful attempt to find a companion for the original man. This individual was simultaneously both male and female. Thus, "He created him" (one body); and later we read "[M]ale and female He created them" (both characteristics). This is not a new understanding. It is an ancient tradition that is found in both the revealed and hidden teachings. The first-century Kabbalistic work *The Bahir* refers to Adam, in section 172 of the book, as androgynous at the time of initial creation. A lengthy discussion, proposing several variations on how the male and female were originally combined, will be found on page 61a of *Berachos*. One proposal is that "a female figure was attached to the male figure." These were both fully formed, and, the woman "needed only to be separated" from the man. The rationalist commentator Rashi says that "He created him with two faces in the first act of creation, and after that He divided them" into male and female. This division was necessary because a suitable companion could not be found among the already existing creatures.

Who were these creatures among whom it was originally thought that a companion could conceivably be found? Some commentators say that among them were others like Adam, the original man. But they had no human intellect. They were only like Adam in body. We have a hint of this in the story of the exile of Cain after he killed Abel. Cain is concerned "that whosoever finds me will slay me" (Genesis 4:14). We are also told that Adam was separated for a time from Eve, and during this time he had relations with other women. Who were these women? Where did they come from? Who or what is it that might find and slay Cain? Various commentators say that there were other man-like creatures. Call these Cro-Magnon, or Neanderthal or something else. The point is that they looked like people, but they were not human. The other point is that much time, on a scientific scale of measurement, could have passed in the meantime. Science does not falsify Torah just because we find fossils that the Torah does not discuss. The Torah does not tell us everything, but only that which is needed for spiritual education. It is not a book of science, or even of history, though there is history and science discussed in Torah.

Why then was it not possible to find a suitable companion for the man, if other manlike creatures were about? The answer is that the companion was to be an *ezer k'negdo*, a helper that can stand against him. This is not a job for an inferior. It is a job for an equal, or even a superior individual. Someone who helps you while standing against you, is an individual who looks out for your best interests even if you do not always realize what they may be. This being needs superior insight to realize what the best interest is. It needs a superior intellect to be able to reason through and explain what the right path is. It needs selfless courage to thwart a friend who is making a mistake. It must be a being of special and superior mental endowments who is also "bone of my bones and flesh of my flesh." That is why one of the animal-like human-shaped beasts would not do. And so a companion was, literally speaking, built (*v'yiben*) out of the original body, to produce male and female, man and woman.

Kabbalistic teachings include heavenly emanations known as *sefirot*. There are ten of these. It is an interesting "coincidence" that the current candidate for the theory that explains everything, known as superstrings, happens to have ten dimensions. Anyway, we are told that when man and woman were separated, the man was left with a greater measure of the emanation of *chakmah*, which is generally translated as wisdom or cognitive logic. The woman, on the other hand, is endowed with a superior measure of the emanation known as *binah*, usually translated as understanding or spiritual insight. When the two combine, we get *daath*, which is knowledge. We can see how the woman is connected to *binah* from the word used in her creation, which is *yiben*. The woman has the highest level of spiritual awareness of all the creatures. One reason is that she was the last to come into existence, and creation progresses from the lesser to the more important in spiritual measure. Thus, she is equipped by virtue of intuitive insight to help the man by standing against him when his reason traps him into false conclusions.

We are told by Lisa Aiken in *To Be A Jewish Woman* that, "The Torah's description of the first woman as a helper has little in common with current conceptions of helping." We are not talking about a servant here. Rather, "The Torah's perspective is that a helper is someone who acts like G-d." And the relationship is reciprocal, because "The same is true for men being able to act as spiritual advisors to their wives." When the relationship is oppositional, and help is not offered or the offered help is refused, she is *k'negdo*, she stands against him. But when the man and woman cooperate, one providing help and the other accepting it, the relationship is that of a selfless helper (*ezer*). It is then that they "form a G-dly partnership" to fulfill the purpose for which they were created.

Apportioned to humanity we have wisdom, understanding and knowledge. With these, we have attributes and powers that no other created being has. We can shape, and we can create, and we can understand and we can act. We are indeed partners in creation. And that is what anthropic thinking and quantum mechanics also claim. Where is the conflict between science and Torah?

Chapter 10

A Theory of Everything

Matter and Structure in the Universe

According to science, everything is made out of the same stuff. All atoms are organized around certain basic, elementary constituents like electrons and protons. All matter is built out of atoms from the various elements. A table, a computer, a fish and a person, it is all the same. We are constructed from the same basic material, and we are organized in accordance with the same organizational laws. The chemical and electrical laws governing the complex interactions in the human body are the same as those involved in a battery or a computer chip or a potato chip. The interesting thing is that the Torah appears to agree, saying that G-d "… formed the man of dust from the ground …" (Genesis 2:7). Is that all we are? A collection of dust? Apparently the answer is yes. But somehow that does not seem right. Even a cow does not behave like a random collection of dust. And a human being certainly does not seem to fit this description. Are we dust, or are we something else? There is a contradiction here that calls for a closer look.

We need to consider matter in both its fundamentally simple form, and in its organized complex form, when exploring how, when and from where this universe and all it contains, including ourselves, comes from. Physicists are not there yet, but it is hoped that all that the universe contains will one day be described by just a few fundamental entities. I could have said elementary particles, but we know that some of the particles don't exactly behave like "particles" ought to behave. Whatever name we use, it is proposed that the universe is constructed from a relatively small number of basic building blocks. Where did these elementary particles, or entities, or whatever, come from? Why exactly these particles and not others? Why do we observe these characteristics and not others? Physics and Torah provide answers. Are the answers in agreement, or not?

The former involves considering each of the building blocks separately. We can also examine these building blocks in organized form. Here, the total is frequently more than the simple sum of the parts. Stars are more than a simple collection of gas. Living organisms are more than a simple collection of atoms. When, how, and why did the basic building blocks get organized in this way? What does it all mean? Is there a meaning at all? What is the organizing principle, or principles, behind these complex structures? How did chaotic simplicity become organized complexity? Torah and science, again, provide answers. Sometimes the answers appear to agree, and sometimes the answers seem to disagree. Can these answers be reconciled?

The structure of matter

The ideal of science is to describe the universe simply as a combination of a few fundamental particles. We can gauge how close we are to this goal by a listing of currently detected, or theoretically predicted, fundamental particles. Here is a partial list, consisting of the better known constituents of the universe: alpha particle, baryon, boson, electron, fermion, graviton, hadron, lepton, meson, neutrino, neutron, photon, positron, proton, quark. It's a long list, and there are others. Furthermore, each particle has an antimatter antiparticle, so we need to double the numbers. In addition, many particles come in a number of versions, such as pi mesons, K mesons, rho mesons, and so forth. And if that is not enough, we also have so called virtual versions of particles. There is the real photon that we can easily detect as in light, and there is the "virtual" photon which is the interaction particle in electromagnetism. This does not look like a list of a "few, simple" particles. Indeed, for a while fundamental particle physicists thought that we might have a problem here. But that is no longer the case. The situation has been sorted out. And while we cannot be certain that all has been discovered, it does look as if the basic structure of matter does indeed consist of a relatively few constituent building blocks, or at least building block categories.

One reason that we have so many fundamental particles is connected to the meaning of the word "particle." Say particle, and most of us have the image of an indivisible little clump of matter. Something solid. Something that has a distinct location in space. But quantum mechanics (QM) spoils that image. Electrons, which we think of as particles of matter, and photons of light, which we think of as waves of energy, have the same particle-wave duality in QM. The difference is not between particle and wave, but between mass for the electron and no mass for the photon. Another distinction that is "obvious" in ordinary experience is between matter and forces. A chair consists of matter, while a push that causes the chair to move, is a force. A chair is made out of particles of matter, while a force is something else. This is not the case in QM. Forces, particles of matter that have mass, such as the electron, and energy waves, such as light quanta, all look and behave in similar ways. The list of particles above has examples from all three categories. Furthermore, some of the names on the list represent a whole class of items, rather than a single individual item. Physicists use a variety of classification schemes to explain what is going on here. Usually the classification is intended to emphasize some deep aspect of meaning, such as symmetry relationships or conserved quantities. We shall mention some of these as necessary, but this is not a treatise on physics. There are available excellent expositions on fundamental physics that the interested reader can examine. Here, the intent is simply to outline a basic level of order and simplicity, so that the reader will see the reason for the final results.

The fundamental particles that produce forces are short-lived virtual particles (though QM fluctuations in the vacuum also produce other virtual particles), also known as messenger particles. These are called *bosons* as a class. Material particles, upon which the forces act, are *fermions*. *Fermions* can be further separated into heavy *baryons* and the lighter *leptons*. *Baryons*, and antimatter *antibaryons*, are created and destroyed in pairs, hence the baryon number is conserved. The nucleus of an atom consists of *baryons*, proton and neutron, held in place by the strong nuclear force. The low-mass electron and muon and no-mass (or maybe extremely low mass) neutrino are the *leptons*. *Leptons* also come in matter and antimatter pairs, and their number is conserved. We also have one additional category designation called *hadrons*. These are particles that participate in the strong nuclear force interaction. This includes baryons which follow a rule known as the Pauli exclusion principle, and mesons, which do not obey this principle. One can get into all sorts of other fine distinctions which are not particularly important for our purposes. Here are the essentials.

There are a number of highly unstable particles that exist for only a very short time. For example, the meson, of which there are several varieties, is an unstable hadron with a mean life of between 10^{-6} and 10^{-16} seconds, depending on the type of meson. Our primary interest here is in the stable matter that sticks around for a while. The negatively charged electron is unconditionally stable. Electrons do not decay into other particles, though electron and positively charged anti-electron (known as positron) pairs can be created or destroyed. The electron is not a hadron, is not affected by the strong nuclear force, and is not a direct constituent of the atomic nucleus, though it does participate via the neutron. Electrons, in QM waves, occupy specific positions around the nucleus, and this is the source of the chemical properties as atoms interact. The positively charged proton is unconditionally stable whether in the nucleus or on its own. It is a heavy hadron with a mass 1836 times that of the electron. The total electrical charge of the universe appears to be balanced at zero. This means that we should have as many total negative-charge electrons as positive-charge protons. Finally, we have the zero, or neutral, electrical-charge neutron. It is a bit more massive than the proton and is unstable, with a half-life of 920 seconds, when not bound into an atomic nucleus. A free neutron will decay into a proton, an electron and a neutrino. But a neutron is stable when bound into the nucleus by the strong force.

The current universe is maintained by four forces. I say "current" because it is almost certain that three of these forces are versions, or manifestations, of what was once one primary force in the very hot young universe. Gravity, which is very much weaker than the other forces, doesn't quite fit yet. But there is good evidence within the logic of grand unified theories (GUT) that the remaining three forces were once one and the same force. Once electricity and magnetism were thought to be different forces. But we know today that they are different aspects of the same electromagnetic force. The weak force, associated with radioactive decay, has now been shown by a Weinberg-Salam theory (for which they received the Nobel prize in 1980) to be part of a higher order force including electromagnetism. This is the so called "electroweak" force. Hence, some physicists no longer refer to four, but rather to three fundamental forces. It is hoped that one day gravity too will join the other forces in a unified field theory, or TOE (theory of everything) which also incorporates behavior on the QM scale. A TOE would explain everything, including forces, matter particles, energy, time and space. Such a scientific description would be fully in accord with Torah which tells us that only one mysterious "something" was created, and all else followed later, built or made or formed out of this initial creation. This is difficult to visualize because these forces are so very different in their current manifestations. One attribute which is the same among all the forces, however, is the fundamental process whereby these act at a distance. All the forces act at a distance by way of virtual messenger particles. The forces differ as to the properties of these particles, but they are all associated with interaction particles.

We experience two of the four forces directly at our distance scale. These are gravity and electromagnetism. Both are long-distance forces without any limit to the distance at which they are effective. Both follow the inverse square law, where the force is reduced proportionally to the square of the distance between interacting objects. The force is reduced by a factor of four when the distance is doubled, by nine when the distance is tripled, and so forth. The force gets very small at long distances, but it never gets to zero. This stands in contrast to the two short-range nuclear forces, known as weak and strong, which are effective over only very short distances.

Gravity is the weakest force by a very large ratio. But it is cumulative. All particles are subjected to this force. And the force never stops, though it is much diminished with distance. The much stronger electromagnetic force has no effect on neutral-charge matter. It takes an electric-charge polarity to create a force interaction. Like charges repel, and opposite polarity-charges attract. This force is enormously stronger than gravity, and it extends indefinitely through space, just like gravity. But most matter is neutral, involving an equal number of positive and negatively-charged particles. That is why only gravity is important at cosmic distances. The cumulative effect of the gravitational force is responsible for many of the phenomena we observe and experience. Stars are crushed inward by gravity to explode as novae and turn into neutron stars where the powerful nuclear binding force is overcome. Gravity appears very powerful at our scale of existence and beyond. But its fundamental messenger particle, the graviton, is so weak that it has not yet been detected. Messenger particles for the other forces, by contrast, have been detected. And the messenger for the electromagnetic force, the photon, is very familiar to us from our direct experience with light. But the mediating particle for electromagnetism is not an ordinary photon. Rather it is a so called "virtual" photon. What is this "virtual" business about?

For a long while there was a puzzle as to how different clumps of matter interact at a distance. How does a stone recognize where the earth is, so that it feels a force towards the earth and falls down and not up? The answer is a force messenger particle called a graviton. Many experiments have been performed to capture, or detect, gravitons. But the graviton is a weakly interacting particle, and none has been caught in the act so far. Not so with the force particle of electromagnetism. How does an electron know that there is a proton somewhere nearby so that the electron feels a force compelling it to change direction and move towards the proton? The electron and proton know about each other through the agency of the exchange and movement of photons. These are not ordinary photons, but short-lived virtual photons. Yet they are photons nevertheless. The force messenger photons swarm about the electron and proton informing, pushing, identifying or whatever word one wants to use, each particle to the other. The result is what we perceive, or measure, as a force between the electrically charged particles. There is also a swarm of gravitons acting between these particles in accordance with the force rules for gravity. But the gravitational force between two protons is weaker than the electrical force between the protons by a ratio of 10^{36}. This is so tiny in comparison to the electromagnetic force that it can be ignored. Unless, of course, the clumps of matter are electrically neutral and there is no electromagnetic force involved. Then, gravity will be the only force operating at a distance. The graviton and photon are massless neutral particles that propagate at the speed of light. They move fast and keep going no matter how far. That is why these are long-range forces.

Next comes the weak nuclear force. This is the force involved in radioactive decay. Every once in a while the stuff that is held together for some reason becomes no longer as tightly held together, and something, a mass-bearing particle like an electron or a massless energy particle like the photon, is emitted. This is a short-distance force because the force-bearing particles involved have mass, so they cannot travel very fast or very far. Three massive vector bosons are responsible for the behavior of the weak force, the W^-, W^+ and Z^0. These are called massive because, unlike the zero mass photon and graviton, these particles carry mass. *Vector* refers to a directional sense, or force component. And *boson* is the general designation for force messenger particles. At a mass equivalent of one hundred billion electron volts, these are truly massive particles. By comparison, the rest mass of the electron is only one half million electron volts. One would think that something so massive would be utterly different from the electromagnetic force particle, the massless photon. But no. It has now been shown that the two are the same at high enough energy levels. The weak nuclear force and the electromagnetic force were the same force in the very hot early universe. Later when things cooled off, the symmetry was broken, and we now have what appear to be two different forces. It is believed that the same idea should apply to the other forces as well. Everything comes from a single, simple force component in the very beginning of the universe. This applies not only to forces, but to hadron matter as well. The fact that everything originates from a single, simple beginning has important implications for a comparison of science and Torah, because Torah makes the same claim.

Finally we have the strong nuclear force. This force is not just strong, it is enormously powerful. It forces the neutron to be fully stable when in the nucleus of an atom. Positively-charged protons repel each other with a powerful electromagnetic force. But this repulsive force is easily overcome once two protons get close enough together to be subjected to the strong force. The force is tremendously powerful, but it is of very short range and confined to the size of the atomic nucleus. Heavy atoms, containing many neutrons and protons, involve sufficiently large nuclei that the strong force cannot be fully effective in the outer boundaries. Not being fully subjected to the strong force, these atomic nuclei become unstable and adjust to a smaller and more stable configuration by radioactive decay. To explain the behavior of this force, a mass-bearing particle with certain properties, which he called a *meson*, was first predicted by physicist Hideki Yukawa. The usual current name for this messenger particle is the *pion*. But this is not the end. It turns out that there is a hidden layer of higher-order simplicity beneath the description of a nucleus made up of protons and neutrons held together by the strong force. This hidden layer consists of *quarks* and *gluons*, and it is important in comparing science to Torah.

Till recently it seemed that the neutron and proton were basic, or fundamental, particles. Well, maybe the basic proton and neutron were not absolutely fundamental because each one of these in certain instances can actually change into the other. Thus, a neutron can decay to a proton plus electron with excess energy carried away by a neutrino. Nonetheless, both appeared to be fundamental particles. But not any more. It is now understood that neutrons and protons, which are stable, and also some unstable particles, such as mesons, are combinations of a more fundamental particle, the *quark*. The force messenger in this arrangement is the *gluon*. Hence, one will frequently find that a description of the atomic nucleus consists of quarks and gluons. It is also common to talk about the early universe as consisting of a quark and gluon plasma out of which ordinary hadron matter appeared as the "frozen" result in a phase transition. Knowing the temperature of this phase transition permits a scientific calculation of a time-dilation ratio. Based on the currently best available information, the time-dilation is a ratio of 6 days versus 15 billion years. Fifteen billion years is the scientifically determined age of the universe, while six days is the creation time from the Torah. Quarks and gluons are important to our story.

The quark was originally proposed on theoretical grounds. The theoretical basis for the existence of the quark is very strong. But there is no longer any need to rely on theory, because we now have solid experimental confirmation that hadron matter consists of combinations of quarks. With the addition of the quark, we have the following basic, or fundamental, particles: the energy-carrying photon; the low-mass leptons, which do not go into the nucleus of the atom and do not feel the strong force but are subjected to the weak force – basically the electron and neutrino; and the hadron-producing quark. There are six types of quarks, known as flavors: up, down, strange, charmed, top and bottom. Experimental verification of the properties of the difficult-to-catch top quark in the late 1990s, has been hailed as final confirmation of the work of Nobel winner Murray Gell-Mann, who first proposed the quark theory. Each quark flavor comes in three versions, called colors: red, white and blue. Stable hadrons, which are baryons, are colorless combinations of three different-colored quarks. The union of two quarks makes an unstable, short-lived meson.

Much more can be said about the properties of fundamental matter in general, and quarks in particular. The electrical charge of quarks comes in units of 1/3 and 2/3, for example. These combine to yield a zero-net-charge neutron and a one-net-charge proton. There is no doubt that quarks exist. But they do not exist as separate, unbound entities at current conditions. The temperature, or energy density of the universe, is much too low now. It would take a tremendously high energy level to break the gluon bond between quarks, even temporarily. That is exactly what scientists will soon attempt to do. The attempt will be made with a machine called the Relativistic Heavy-Ion Collider (RHIC), which is scheduled to go into operation at the Brookhaven National Laboratory. Streams of heavy ions, such as gold, will be accelerated to near light-speed, causing them to crash into each other. It is expected that this will bring about conditions for the existence of a mixture of free quarks and gluons. The early universe was dominated by energy quanta, which is a form of intense light. There were also material particles present, but these behaved the same as energy quanta at these enormous temperatures. This involved the low-mass electron, the almost no-mass neutrino and their antimatter counterparts, as well as heavier elementary quarks. These matter and antimatter pairs were constantly created and destroyed in a brew of intense light, energy and pressure. But the universe was expanding and cooling. Soon, the temperature got cool enough for ordinary hadron matter to "freeze" out of the quark-gluon plasma, and stay solid. This is known as the phase transition temperature. An analogy is the temperature where liquid water undergoes a phase transition to solid ice. The RHIC project will attempt to determine this temperature which is predicted to be in the range of 10^{13} degrees.

The Torah also has a light-and-energy-dominated era, from the moment of "Let there be light" (Genesis 1:3), till solid matter is brought into existence. This would correspond to the quark-to-hadron phase transition point. This is an important milestone because we learn, as is discussed by Nachmanides, that this is the instant where time can be dealt with in an ordinary fashion. Counting time from this moment, comparing the phase transition temperature to the current cooled remnant known to be 2.73 degrees, and using relativistic time-dilation theory, we can compute a time ratio for the first six days of the universe looking forward in time compared to the current age of the universe looking backwards in time. This is discussed elsewhere in this book and with specific detail in Dr. Schroeder's *The Science of G-d.* I have no way of knowing what temperature the RHIC experiments will show. But I would very much like it to be $9.7 x 10^{12}$. Such a temperature would yield a time dilation ratio where 6 days in the beginning is equivalent to 14.2 billion years (BY) at this time. I like this number because 14.23 BY is my favorite Torah-based calculation for the age of the universe.

All right, so there might be some connection between the existence of quarks and Torah. What about the rest of the stuff? There has to be a connection. That is the premise of these essays. Everything that was created has a purpose. Hence, there must be a connection. Some of the relationships are known to us, and some are not. Twenty years ago nobody dreamed about quarks. The connection to the teaching of the Torah would not have been clear to us. Now science has caught up, and we can understand some things. Who knows what will yet become clear in the future? Let us take a look at the mysterious, and seemingly useless, neutrino in this context.

Why do we need neutrinos?

Photons to carry energy, electrons for making chemistry upon which life depends and quarks to make solid matter out of which we and all else are built – these three basic components are all the universe needs. Yes, we need virtual messenger particles to make forces, and there are all sorts of short-lived entities that seem to do something, but these are just a way to help the stuff that counts to do its job. The virtual and short-lived particles are, in a sense of speaking, not even real. These are sort of like the frosting on the cake. We cannot say the same for the neutrino. It is one of the most durable and stable particles around. To capture, affect or destroy a neutrino is an almost impossible job. It would take light years worth of lead shielding to stop a neutrino. The neutrino is here, but what is it good for? Yes, it does perform a useful function to carry away excess energy when an atom undergoes radioactive decay. But there are other ways in which this bookkeeping trick could be handled. Instead of emitting an electron and neutrino when a neutron turns into a proton, how about emitting the electron at a higher velocity to carry away the excess energy? Why does the universe need a neutrino that hardly interacts with matter in any way? It is a mystery.

Neutrinos were first hypothesized in the early 1930s by Wolfgang Pauli as a way to account for the missing energy in radioactive decay. But the particle was not detected till the 1950s, when vast numbers of concentrated neutrinos first became available from nuclear reactors. The interaction of the neutrino with matter is almost nonexistent, but not zero. Concentrate enough neutrinos onto a detector and you will catch a few. There are several different kinds of neutrinos and their antiparticle versions, the antineutrino. All are characterized by almost zero mass and near light-speed velocity. The neutrino is not affected by anything except the force of gravity and the weak force involved in the radioactive decay process. No one is sure how many neutrinos there are in the universe. But there are theoretical grounds to indicate that there are as many neutrinos as there are photons.

Neutrinos were created in the original big bang process of the creation of the universe, and new ones are constantly streaming out of stars and radioactive decay. The mass of the neutrino is tiny, if any, so they do not contribute appreciably to the content of the universe. They move at the speed of light if massless, or at nearly that speed if there is some tiny mass associated with the neutrino, as recent tests indicate. The neutrino is like a photon with respect to mass and velocity. But a photon serves an essential function as it interacts with matter. We see by means of photons. We send radio signals through photons. We transmit energy through photons. Not so with neutrinos, because they do not interact with anything. The sun shines down on the earth, and we see the light photons. At night we see no photons because the sun is hidden from view. But day or night is all the same for neutrinos because the earth is transparent to them. A light year's width of lead is for an electron-neutrino like a pane of glass for a light-photon. But what is this thing good for?

Possibly there is no mystery involved. We do need a mechanism to keep the energy before and after balanced in the radioactive transformation process. The neutrino provides a simple way to do this. Alternatively, the neutrino will perhaps explain some very deep insight into science or Torah one day. Too bad that this matter has to remain hidden at this time. But possibly the secret has already been revealed to us in the supernova process.

In 1987, the neutrino flux from a supernova in the Large Magellanic Cloud, 170,000 light years away, reached the earth. A total of 22 antineutrinos were detected (11 in Japan, 8 in the United States and 3 in the Soviet Union). It is estimated that the neutrino flux emitted by the collapsing star was 10^{58} neutrinos. Of these, 10^{11} neutrinos passed through every square centimeter on earth, and about 10^{14} neutrinos passed through every person. Twenty-two detector interceptions out of such a large number shows how weakly the neutrino interacts with matter. This is important for our health, because no one could survive being hit by 10^{14} interacting particles. It is good that neutrinos don't interact much with matter, otherwise they would kill us. But not having any neutrinos would also not kill us. So why neutrinos? Possibly because neutrinos are a necessary factor in causing a supernova explosion. And the supernova is a necessary factor in the spreading of carbon. And carbon is essential to our existence. So in this respect, it does seem as if we would not exist if there were no neutrinos. But a flood of strongly interacting neutrinos would kill us. Does this perhaps explain the purpose of this mysterious weakly interacting particle? You decide.

The enormous gravitational pressure of a collapsing star turns the iron core into pure neutrons. This means that protons become neutrons, releasing a neutrino in the process. This is where most of the neutrinos from the star come from. Neutrinos are weakly interacting particles. But a neutron ball is not ordinary matter. It is tremendously dense, compressed matter that even the elusive neutrino cannot easily penetrate. The neutrinos hit the neutron core of the star and bounce back with great force. The star explodes spewing everything except the neutron core, all over space. Among the debris of the star is carbon, which was cooked in the nuclear furnace of the star. A billion years later this carbon combines with other cosmic dust to form a new generation star and new generation planets where carbon is a common ingredient. Carbon is part of the planet; it is part of the "dust of the earth" from which man was fashioned. Having performed the job of exploding the star, the neutrinos speed harmlessly on their way. In 1987, the creature called man catches 22 neutrinos in traps devised by his ingenuity – an ingenuity which is uniquely ours as part of being created "in the image" of Him who brought it all into being.

The organization of matter

It appears that the universe consists of four-dimensional space-time which contains leptons (electron, neutrino), quarks (which make up nuclear particles, such as the proton), force messenger particles (such as graviton and gluon) and the energy wave particle, the photon. That seems simple enough. Everything consists of these fundamental energy, force and particle components. They are to the universe what the ten numerals from zero to nine are to mathematics. They are to the universe what the letters of the alphabet are to writing or literature. So of course there is more to say here on the subject of the organization of matter. We can begin by distinguishing between two basic categories. How did existing organized structures come about, and how do these structures maintain their existence? Form and continuity of structure depends upon behavior. The behavior of some existing structures is relatively easy to explain. Take a star, for instance. The nuclear processes operating within the star are well understood today. We understand how the outward pressure of the burning nuclear fires is balanced by the inward pressure of gravitational attraction to provide a stable balanced environment. We understand why and how the star explodes as a nova when this balance is destroyed when there is no more nuclear fuel to burn. We understand why and how certain stars will explode in a more violent fashion as a supernova. We understand how some rather obscure processes, such as neutrino flux pressure, are necessary ingredients in the supernova explosion. We understand this and much more. We understand and marvel at the intricate and delicate balance that makes it all come out just right. We wonder why it is this way and not some other way. Is it possible to get something so well attuned to our existence by sheer accident? Maybe yes, and maybe no. Whatever your choice, we have a "natural" process explanation for how it works. We may choose to look for other causes, but we are not compelled to do so. The fact that everything looks "natural" is fully in accord with Torah. Time and again, Torah tells us that the world is made to look natural and understandable to us. Only in special cases, and only with sophisticated effort, can we discern the true Power that is in control of the universe. Mostly we are told that something was fashioned, structured or built using existing materials and in accordance with rules or laws that we have the intellect to comprehend. No wonder it seems "natural," and in no need of additional explanation.

But not everything is as easy to understand as the behavior of a star. How about ourselves? The human genome project is making great progress. Do we know what makes a human being now that we have all the genes deciphered? Some say yes. But the Torah says no. Even if we knew the full genetic code. Even if we knew all the chemical and electrical interactions in the body. Even if we could explain the function and behavior of every molecule and every atom. Even if we could explain the behavior of every physical interaction. The Torah would still say no, even if we could answer all the previous questions and any other that could be asked about the human body. Not because the Torah disagrees that these can be ultimately understood. On the contrary, the Torah assures us that these matters and much, much more are in principle within our grasp. Anything that was fashioned and built can be ultimately understood by the special being called man. But a body, fashioned out of existing materials, is only one component of what we are. The second component is pure spirit. Science cannot deal with this. But science does not have to, because science denies that such an entity exists.

Based on the above, we can eliminate virtually all aspects of current existence from a discussion that seeks to explore disagreements between science and Torah. This is because some of the simpler matters, such as a star, can be fairly well explained by current science, and while more complicated entities, such as the human body, cannot yet be explained, science shows great promise, and the Torah indicates that we should be able to do it. The remaining aspect of existence is in the spiritual, rather than physical, realm. Science cannot explain it. But science has no need to explain that which science does not recognize as existing. It is not intellectually satisfying to eliminate so much material with the simple statement that there is no need to explore it. Indeed, there are numerous works on philosophy and theology that deal with these matters. But it is not an issue between science and Torah.

Besides, we need to concentrate our energies on a more important item. There is still the matter of how everything that now exists came to be this way. Here, science and Torah have much to say to each other.

Let us first take the better known, and more exciting, case of the human body. How did it come about? Science points to evolution. Disordered atoms, molecules and proteins slowly acquire the complex ordered structure of the human body. Order out of disorder. Does this not violate the second law of thermodynamics, which requires that disorder increase as a function of time? No. This principle applies only to a closed system. The increase in order, for the living organism, is more than offset by the additional disorder of the outside environment. Evolution does not violate basic physical law. The essential debate with evolution, as discussed in detail in earlier chapters, takes place in the field of probability. Can it possibly have happened within the available time? There are those who "prove" that it can, and there are those who "prove" that it is impossible. It all depends on the choice of basic assumptions. The reader should by now understand where this writer stands on this matter.

Another problem with evolution is that certain predictions and necessary conditions do not fit the actual results. Why or how did life appear so early after the formation of the Earth, contrary to predictions from evolution? Why is it impossible to find some essential intermediary forms? Why do living organisms show a very high level of stability over long periods of time, and why do new ones appear so suddenly? These are serious questions. But that does not kill evolution, *per se*. It could just be that the science is incomplete, as is quite common with any scientific theory. Give us time and we will straighten it out, say the true believers. It is not possible to prove that science will not find answers. All we can say is that the above questions, and many others, are easily answered by Torah. Torah fits the results of detailed observation. Furthermore, Torah also fits the general idea that mankind is fashioned out of the ordinary matter of this world. The Torah states directly that the human body was fashioned from the same materials as the dust of the earth. Evolution of living organisms is an emotionally charged topic. But it is, in fact, not a serious obstacle in reconciling Torah with science. Evolution of the universe is actually of greater significance. There are a number of scientific issues or problems when dealing with the evolution of the universe, starting with the initial big bang explosion. Any extensive work on cosmology will allude to various issues, such as those discussed below.

We know that disorder, measured by a quantity called entropy, must increase with time in a closed system. This is a fundamental understanding of physics. It is one of the indicators that time has a preferred direction, and is known as the thermodynamic arrow of time. Where did all the current order come from? I previously noted that there is nothing remarkable about the order of living organisms, thermodynamically speaking, because the increase in order, for the organism, is more than offset by an increase in disorder of the total environment. This applies not only to living organisms but to the other complex, ordered systems within the Earth's biosphere. But that presupposes that the outside environment is sufficiently ordered that there is room for disorder to increase. It would appear that, somehow or other, the universe as a whole had to be more ordered in the past than it is today. Where did this order come from? Where is this hidden order stored? There is a surprisingly simple answer to that question.

The answer is that hydrogen is a repository of order. Here is why.

Order involves differences, while disorder makes all parts of a system all the same on the average. A box with air molecules evenly distributed with the same pressure everywhere is disordered. The same box, experiencing a statistically rare (should we say, absolutely super-rare) fluctuation, with most of the air to one side and the other at much lower pressure, exhibits a higher level of order. A glass of clear water with a drop of ink injected to one side is in a state of order. Given time, the ink will diffuse uniformly throughout the water, and the mixture will now be in a state of disorder. The same applies to heat. Heat flows from the higher temperature to the lower temperature. This involves a transfer of energy, and work can be performed in this manner. A system that permits energy to flow and work to be done is in a higher state of order than one where energy cannot be transferred or work be done. Order on earth feeds on the input of sunlight, which permits energy to be transferred and work to be performed. Sunlight in turn comes from the nuclear burning of hydrogen as it is slowly converted to the nuclear residue we call iron. About three quarters of the early universe was hydrogen. This is the source of the order. Apparently, all we have to do is explain why everything started with such an abundance of hydrogen, and we have explained the basis for the current order in the universe.

But what is apparent on the surface, is not so apparent upon closer examination. The scientific explanation as to why so much of the early universe consisted of hydrogen is simultaneously both simple and mysterious. It is simple because we can show how it happened. It is mysterious because we do not know why it happened this way instead of some other way. The mystery relates to the reason that certain universal parameters have particular values. For example, why are the particle masses what they are? Where do these numbers come from, and how are these numbers interconnected? Are these numbers interconnected, so that if we know one, we also know the others? Why is it this way? As stated before, there is a simple explanation, but it does not satisfy. The simple explanation is that we would likely not exist if these values and relationships were other than they are. Why was there exactly the amount of hydrogen that there was in the early universe? Science can compute the result based on the values and interrelationships between particles and forces. It is not all that simple, actually, but the matter is well understood. What we don't know is why the values are as they are in the first place.

In 1995 the predicted top quark was detected and measured to have a mass of 180 billion electron volts. By comparison, the proton mass is just under one billion electron volts. Why? Nobody knows. Scientists knew that there must be a very massive quark. But its mass could not be predicted. In fact, the value was expected to be about 120 billion electron volts, and the result was a surprise. We don't know that anything terrible would happen to us if the mass of the top quark were just a bit different. We do know that if it were seriously different we would not be around. So why is the value what it is? What forced it to be this way? Is it possible that a small change would in fact be catastrophic, but we do not yet understand quarks enough to see this? We do not know.

We have had more time to understand the electromagnetic interaction better, and we know that a fairly small change would be most unpleasant. What exactly would happen if the electromagnetic strength were only 0.5% different? The fusion reaction in the sun is dependent on the ninth power of the electromagnetic force strength. This turns a 0.5% change in this force into a 4.5% change in energy output from the sun. And that is enough to mess everything up on Earth. There are about eighteen basic parameters whose values appear to be arbitrary. Yet everything ultimately depends upon these values. Why is it this way? The universe also depends on six critical ratios. Martin Rees tells us in *Just Six Numbers* that "if any one of them were to be 'untuned', there would be no stars and no life." But the values of these numbers appear to be totally arbitrary, and "we cannot predict any of them from the values of the others." Our existence depends on these numbers. But nobody understands where they come from. Something may come up in the future, but right now we can site the anthropic principle which requires the universe to be friendly to intelligent life, or we can cite the Torah. Take your choice.

The former deals with the very small scale, on an atomic level. Now let's look at issues on a large scale, where gravity is the controlling force. There are many fundamental problems here for science. All relate to the mode whereby the current universe acquired complex organization out of the disorganized chaos. The Torah has a simple answer. First came the creation (*barah*) of space-time and matter-energy. The remainder of the six days, which we call creation, was not really creation. It was the organization of the existing universe into a coherent structure. We are told that the human soul was (*barah*) created. The rest was shaped, built and fashioned. It is a simple statement that nicely meets the requirements of the scientific principle of simplicity, known as Occam's razor. Science also has an answer, in the modification to the big bang theory, known as *inflation*. Inflation works very well in explaining and predicting results. Inflation is not directly identified in the Torah. But the process is discussed. There are a number of hidden hints and even some open discussion. The connection is not all that difficult to find once we know what to look for.

Let's begin with the problem of the uniformity of the observed universe and the event horizon. The universe had a specific beginning, say 15 BY (billion years) ago. The fastest that a signal can travel is at the speed of light. Hence, we can only observe, or know what happened, to a distance that light can travel in 15 BY. This is our event horizon[Δ]. We cannot possibly observe anything beyond that point. But we can look 15 BY in two opposite directions. The two most distant points that we observe are farther than 15 BY from each other. Hence we can say with some certainty that there could not be any interaction or communication between these. There simply has not been enough time since the universe began for information to travel that distance. Each point is beyond the event horizon of the other. The event horizon distance depends on the age of the universe, and was exceedingly small in the early universe. The early universe had a huge number (as many as 10^{80} in the very beginning) of uncoupled regions that could not communicate with each other. Each of these could develop in its own independent way. But all parts of the universe that we observe follow the same laws and have similar matter and energy distributions. How did the many independent regions of the early universe communicate and interact so as to influence each other towards a similar final result?

A connected problem is the high degree of isotropy, uniformity with direction, of the universe. An explosion, no matter how well balanced, will have some force and matter distribution variations with direction. This would apply even when all parts of the explosion are within communication distance of each other. Yet measurements of the cosmic background radiation show agreement to better than one part in a thousand. A random "big bang" explosion would not be this uniform.

The evolution of the current universe, consisting of orderly structures of stars, planets and galaxies is even more remarkable when we consider the meaning of order and disorder in a gravitational environment. We previously addressed the difference between order and disorder as a question of energy flow from the more ordered to the less ordered. Let us look at the same question as a matter of probability of occurrence. Highly ordered systems have a low probability of appearing by chance, while highly disordered arrangements have a high probability of occurring by chance. Take the example of the drop of ink in water. The ink and water combination is in a state of high disorder when the ink is uniformly diffused throughout the water. The system is orderly when the ink is concentrated into one dense droplet. The concentrated ink condition can happen if you wait enough billions of years for that rare combination to occur. Not so for the diffused ink situation. You wait a few minutes and you have it. Not only that, but the appearance of the slightly inky-colored water never seems to change. Molecules of ink and water move about all the time. But the total picture does not change once the maximum condition of disorder is reached. Why? Because the number of possible combinations of ink and water molecule arrangements is enormously greater for the diffused-ink condition than for the ink-in-one-spot condition. The probability of finding a distributed-ink arrangement is large, very large, in comparison to any other arrangement. So what does all this have to do with gravity? Quite a bit, actually.

Order and disorder in a gravity-force environment is counterintuitive. What we would normally consider order is disorder, and vice versa. Take the whole universe with matter distributed throughout. Let us ignore the electromagnetic force on the basis that, on the average, we have a neutral electric charge situation. All there is, outside the atomic nucleus, is the force of gravity. This force acts between all the items in the universe. Everything attracts everything else. The closer two clumps of matter are to each other, the greater the attraction. Also, the more massive the clumps of matter, the greater the attraction. Soon atoms join together into larger and larger combinations. These attract each other and join together. And these further attract each other and join together. This is basically why we have planets and stars and clusters of stars which we call galaxies. These entities would collapse and become tiny and super-dense in the absence of other internal forces. Fortunately for us, the internal electrical and nuclear forces of planets and stars keep these from collapsing. But not for long.

Pile enough matter on, and the cumulative effect of gravity will overwhelm everything. Eventually the internal structure will collapse into a so called "black hole." The gravitational field of a black hole is so intense that not even light can escape. That is why it is called a black hole. The interesting point for us is that analysis of a random distribution of gravitating matter shows an overwhelming tendency to form black holes rather than stars or clouds of gas. Strange as it seems, a universe full of black holes is at a higher state of disorder than one of stars, planets and clouds of gas. This counterintuitive idea was verified when Stephen Hawking and Jacob Bekenstein showed that the disorder of black holes, as measured by entropy, is very much greater than for a star of the same mass. The entropy ratio between our sun and an equally massive black hole at a mass of $2x10^{33}$ gm, is 10^{18} times. This is a huge ratio. A theory by Dr. Mario Rabinowitz suggests that the universe must be full of what he calls, "little black holes" with a density of up to 10^{30} black holes per cubic light year of space, for a universe with a 15 billion light-year radius in size. But no one has been able to find any of these black holes. Why is the universe behaving this way?

There are two basic answers to the above, and other similar questions. One answer is provided by Torah. First was the initial creation which took but a moment. The remainder of the six days was used to shape the originally created disordered material into an ordered universe. We would call this shaping a miracle. But it does not violate the second law of thermodynamics. Remember that this law only applies to a closed system, and the second law does not apply to forces coming from outside the universe. The second answer is a scientific theory called *inflation*. Inflation answers these and other questions, and predicts results which have since been verified. Inflationary cosmology is now fully integrated within the standard big bang model. Does that mean that the Torah is now irrelevant? Of course not. And let me give three reasons.

Except in a few instances involving the word *barah*, it is intended that creation should look natural. We have been granted unique powers, by being created "in the image," to unravel and understand the shaping process of creation. And possibly, we might even do some such creating ourselves. Looking "natural" is not contrary to

Torah. A second reason is that science proposes a process called inflation which explains "how" certain things came about. But it does not explain "why" it happened that way. Alan Guth, who first proposed the inflation idea, states on page 276 of his book *The Inflationary Universe* that one day "the existence and history of the universe could be explained by the underlying laws of nature." But that does not really end the questions. Because, as Guth states, "What is it that determined the laws of physics?" Torah, on the other hand, deals mostly with "why," though some hints on the "how" are also provided. Finally, and perhaps most compelling, is the fact that the inflation process fits in nicely with the hints of "how" that the Torah does provide. This situation is an excellent example in the mind of this writer, of the complementary nature of Torah and science. Each provides clues to the other, and each leads towards a deeper appreciation of the other.

Δ The event horizon distance for an expanding universe is actually much more than the age of the universe, as explained in the footnote to Chapter 2. But this does not affect the final conclusion. Here I am using the horizon distance for a static universe, which makes for a much simpler calculation.

Chapter 11

In The Beginning

Was Everything

The first words of the Torah are: *Breishis barah.* In the beginning created. What was created in the beginning? Time and space, which together make space-time and the basic, primordial matter out of which all else is built, came about in the initial physical creation. All else, excepting the human soul, which is a spiritual and not a physical creation, was shaped or built out of the initial creation. Says Nachmanides on pages 22-25 of the Chavel translation of his commentary on Genesis, "Now with this creation, which was like a very small point having no substance, everything in the heavens and on the earth was created." The point had no substance, that is no size or dimensions, but this creation contained all the matter-energy that was to make up the universe. This matter was not the ordinary matter that we now have in the universe. Rather it was "a very thin substance devoid of corporeality but having a power of potency …. This was the primary matter created by G-d" and after this "He did not create anything, but He formed and made things with it."

All of this, the space-time point and the primordial matter, was "brought forth from total and absolute nothing." Not just "nothing" which we might call a vacuum, but "absolute and total nothing." This nothing is less than a vacuum, given that space and time also were created. The size of this creation was a "point having no substance," what science would call a singularity, or something close to it. And then this insubstantial point expanded, and the universe we now have came into being. The initial energy and primordial matter changed form as the size of the universe increased and became the ordinary hadron matter that we are all familiar with. Science tells us that the initial expansion of the fledgling universe was very rapid, but only for a brief moment. Then the rapid expansion suddenly stopped, and the expansion proceeded in the relatively leisurely pace that we know today. The initially rapid expansion is called *inflation*. How do the descriptions of inflation in science and Torah compare?

Inflation

As John Gribbin and Martin Rees put it on page 26 of their book, *Cosmic Coincidences*, the "fine-tuning of the Universe" for life seems to be "of exquisite precision." Details of this precision, and the utterly impossible odds against this happening by chance are discussed in previous chapters. The implications of this precision have for a long time made scientists uneasy. But now science has found a way to explain many of these odds by way of a "natural" result. And chances are that others of the impossible coincidences will also one day be understood as part of the natural order of the universe. Scientists are diligently looking to discover a theory of everything, and Torah assures us that they will succeed. Does this not harm the cause of Torah? No. Time and again Torah tells us, as is noted in earlier essays, that all is arranged in a manner that can be explained as part of nature. In fact, it is essential that the Divine be hidden within nature. Why?

The whole book, *Worldmask*, by Rabbi Akiva Tatz is devoted to answering, and exploring the implications of, the above question. He asks rhetorically, "… why is there a world of nature? Why is there a world of natural-seeming cause-and-effect which certainly appears to be independent and self-maintaining? For what reason does the Creator set up a world that appears to hide him?" His answer is that "the most easily understood reason is in order to provide man with his most important possession: free will." One indicator of this is to note that "miracles reveal that nature is a mask. Behind that mask, there is none other than Hashem Himself." The fact that the world is so arranged that Hashem is hidden from us, is not just a philosophical concept. It also affects us on a practical level. Here is how Lisa Aiken puts it in her practical book *Why Me, G-d: A Jewish Guide for Coping with Suffering*: "The Hebrew word *olam* means world, but it comes from the root *he'elem*, meaning hidden. G-d made the physical world a place where His Presence would be hidden." But while the Cause is hidden in "nature," the natural-seeming explanation for the results is contained within the Torah, if we but know how to look.

We discussed previously the puzzle of a flat universe and the apparent uniformity between distantly separated sections of the universe that should not have had time to affect each other. These, and a number of other probabilistic difficulties, are resolved by the inflationary hypothesis of the big bang theory. And yes, a clue to the rapid expansion of the universe, known as inflation, is indeed found in the Torah.

The idea of an inflationary expansion of the universe was first formally proposed by Alan Guth of MIT in 1980. The expansion started within the precursor of the ordinary vacuum space, known as the "false vacuum." Now twenty years later, the original version of inflationary expansion has been considerably modified. According to Guth, in his book *The Inflationary Universe*, there are at least 50 variations vying for supremacy. These will eventually be sorted out on the basis of measurements. The basic elements, though, are pretty much agreed on, on the basis of current measurement results and theoretical consistency. The inflationary period began very early, before 10^{-37} seconds, and possibly even at the Planck time of 10^{-43} seconds after the big bang. The exponential expansion was very rapid, with an expansion factor of at least e^{60}, which is a rate of 10^{26} in decimal notation. Guth says that there were at least 100 doublings (10^{30} in decimal notation) of the size of the universe in less than 10^{-35} seconds. Each doubling lasted no more than 10^{-37} seconds. This is an impossible to visualize rate of increase. Suffice it to say that the entire universe could be significantly larger than the portion we can observe at a 15-billion-light-year distance. Guth estimates a ratio of 10^{23}, on page 186 of his book. Acceptance of the anthropic principle is much strengthened by this suggestion.

There is currently widespread agreement on the following sequence of events: The grand symmetry of a unified force broke apart by 10^{-35} seconds, when the temperature was about 10^{29} degrees. Inflationary expansion ceased soon thereafter, when the temperature stopped its steep drop as the rapid expansion came to an end, and the braking action reheated the universe to about 10^{27} degrees. The temperature continued to drop to a bit over 10^{15} degrees as the universe proceeded to expand at the current, comparatively leisurely rate till the electroweak force broke apart into electromagnetism and the weak nuclear force at the ripe old universe age of about 10^{-11} seconds.

Today, 15 billion years later, the temperature of the universe is near 3 degrees, and we are here wondering what this is all about. And one of the things that we are curious to know is why everything happened at just this time scale and temperature range to give us a universe perfectly suited to beings like us.

The inflationary high rate of expansion was brief, but the philosophical implications for scientists is enormous. According to Stephen Hawking on page 127 of his *A Brief History of Time*, without inflation the only logical explanation for the basic structure of the universe is as an "act of a G-d who intended to create beings like us." But inflationary expansion of the early universe removes many of the obstacles to a "natural" explanation of how it all came about.

Scientists like inflation because it gets rid of a number of unpleasant factors such as the need to postulate deliberate design. Still, it is only a theory on top of another theory (the big bang). But these days inflation is on a much more solid footing than it used to be, based on measurements reported in early 2000. You do not need to track down the scientific report in *Nature* 404, 955 (2000), because the May 8, 2000, issue of *Newsweek* magazine contains a nice summary of the results. This involves highly detailed measurements of the structure of the cosmic microwave background radiation taken from a balloon at a 23-mile height. You will find the stunning images and explanation of results at the project BOOMERanG (Balloon Observations Of Millimetric Extragalactic Radiation and Geomagnetics) web site at URL www.physics.ucsb.edu/~boomerang/press_images/index.html. The NASA news release dated April 26, 2000, announces that "An international team of cosmologists has released the first detailed images of the universe in its infancy .... The images reveal hundreds of complex regions that are visible as tiny variations – typically only 100 millionth of a degree Celsius (0.0001C) – in the temperature of the CMB ..." (cosmic microwave background radiation). The images are astonishing in their complexity and beauty. Hopefully this web site will still be there when you read this, so you can see for yourself.

Analyses of the above results verify that the early universe behaves in accordance with predictions from the inflationary hypothesis. It is quite clear that the universe is flat with a total density parameter equal to one. Furthermore, results from BOOMERanG interact with, and confirm, the results obtained from measurements of Type Ia supernova, as is noted in my earlier chapter on the age of the universe. We now have strong evidence that the vacuum energy density exceeds the matter density by more than a 2:1 ratio, and the age of the universe appears to be very close to 14 BY. This is exciting news for me, as these results fit perfectly with predictions from Torah. I am particularly happy with the 14 BY number, because my favorite age calculation from Torah shows 14.2 BY.

Inflation cosmology relates to a one-time-only event. It happened once in the history of the universe, never to be repeated. As noted, inflation is a theory on top of a theory (big bang). It is a theory proposed specifically as an explanation of certain observed results. Inflation postulates behavior that is contrary to the currently ongoing laws of physics. Does not the above point to what we commonly call a miracle? Namely, a unique, singular event that does not conform to the standard laws of the universe? Yes, but Torah tells us that science will find a way to assimilate this into the natural order. Indeed, the Torah proposes that all observed physical phenomena, no matter how unique, will one day be explained through a natural-seeming theory. As Rabbi Dessler has told us, "Even if they were shown open miracles ..., they would invent an entirely new theory." And "The miraculous would become nature." The Torah calls for a scientific explanation of phenomena. This is all phenomena, even the seemingly miraculous. But these same phenomena are also described in the Torah. And the Torah description predates science by millennia. Let's take a look at the Torah and the rapid expansion of the early universe.

Inflation in Torah

One should not be surprised that the Bible does not directly describe this unique event, given that all of creation is presented in just a few terse sentences. There is, however, a unique, one-time-only happening as we are told in Genesis 1:2 that "the Divine Presence hovered ...." Schroeder analyzes this statement and identifies it, in his writings, as the initiation of the inflationary process. We also have another event, in Genesis 1:6, involving the "firmament" which is connected to the conclusion of the inflationary time frame by commentary from Rashi and Nachmanides, and the Talmud Tractate *Chagigah*.

Rashi notes that the heavens were created in a "flexible state" and only became fixed in the final form when "He said" ... "Let there be a firmament." Rashi arrives at this conclusion by analysis of Job 26:11. The significance of the above is in the originally "flexible" – changeable, malleable, dare we say expanding – state of the heavens, and the reference to Job which is also given in *Chagigah*. The significance of Nachmanides is in two parts – what he says and what he chooses to conceal. It is significant that

Nachmanides explains that the originally insubstantial substance, "created from nothing," should now become a firmament of real substance. This is also what inflation proposes as the original quantum fluctuation substance went through a phase transition to become the permanent stuff of this universe. He then tells us that there is more involved here but "to give the interpretation is forbidden." What additional information is Nachmanides concealing here? He does not say.

We then come to *Chagigah* page 12a, where the expansion of the universe is clearly described. Not only that, but the manner in which this expansion was stopped is also given to us. Rav Yehudah speaking in the name of Rav says that when the world was created, "it was expanding continuously." This process did not continue but was arrested at the appropriate time, which Rav Yehudah derives from an analysis of Job 26:11. Resh Lakish then tells us that the cessation of the process is connected to the Divine name, El Shaddai, because, "I am the One Who [*sha*] told the world, enough [*dai*]." Furthermore, speaking of this Name, which first appears in Genesis 17:1, Nachmanides says, "with it are done hidden miracles." Rabbi Dr. Chavel comments on this matter on page 215 of his English translation and commentary on Nachmanides. He explains that "hidden" miracles are ones that people can deny and claim to be part of the natural order. And why was it necessary to stop the expansion at that moment and not some other time? There are a number of comments on this, all providing a similar answer. The rapid expansion had to stop so that the heavens and earth would "stand together" The meaning of this phrase is explained in *Chagigah* 12a as, "they will not slip away from each other." Thus, we read in Isaiah 48:13 that after the heavens and earth were put in place, "I call to them and they stand together." "I call to them" stops the rapid expansion, as explained by Resh Lakish, "and they stand together" so that the universe will not separate into useless bits of matter. Scientific theory also tells us that the initial rate of expansion was perfectly balanced to produce a universe where we can exist.

Chagigah clearly tells us that there was an expansion of the beginning universe. It also tells us that the expansion stopped just in time to give us our kind of universe. We also learn that this was accomplished through a hidden miracle. But does the Torah describe an inflationary expansion? I do not know. The Hebrew statement is that the universe got wider – *marchiv veholech. Marchiv* means to widen, expand or dilate. *Holech* usually means to go. In this construction, the phrase would imply an ongoing expansion, and the English rendition in the Art Scroll translation has it as, "it was expanding continuously." The expansion was continuous as in smooth rather than a start-stop-and-start-again manner. But it was not of long duration, and we are clearly informed that the expansion was soon arrested. We are also informed how this expansion was stopped. There are many other aspects to this issue that I am not discussing, because these fall into the hidden knowledge which I am not qualified to discuss. We are told, for instance, that the expansion was like two threads unraveling in a loom. We know from basic Kaballistic discourse that these threads are analogues of the primordial matter out of which the heavens and the earth were constructed. Possibly there is some hidden understanding in the loom analogy respecting the acceleration, or lack thereof, of this expansion. But I have no knowledge of it. All I can say is that the Torah clearly indicates that the early universe underwent a brief period of expansion – and that the nature of this expansion was that of a hidden miracle, which means that it can be explained by means of natural law. And that the substance of the universe was transformed by this process from being insubstantial to firm, natural matter. As I said, inflation does not directly appear in the Torah, but we do have hints of it.

Guth, on page 171 of his book, provides an illustrative analogy to explain the repulsive force that drives the inflationary expansion of the false vacuum. He shows a cylinder filled with false vacuum and a hand pulling on the piston to increase the volume of the cylinder. He then asks, rhetorically, where the energy to pull on the cylinder comes from. His answer is that it comes from the "hand that moves the piston outward." I am sure that Guth means nothing special by this phrase. But it conjures up a powerful image for someone familiar with the analogy involving the hand of Hashem used in Torah. There is no fundamental conflict here between science and Torah, because Torah supports the idea of a rapidly expanding universe, and it also supports the idea of a natural-seeming explanation for the expansion. A deeper analysis shows that there is more than that involved here. But we need to consider a few more aspects of the beginning of the universe before taking up a more detailed comparison of Torah and science on the matter of inflation.

In the beginning was "absolute and total nothing"

Where did it all come from? It came from nothing as a QM-based fluctuation within, what is known as the "false vacuum." The false vacuum is a nothing that is less than the nothing of empty space. To understand how we can get to a nothing which is less than nothing, it is necessary to explain some things about the mysterious and counter-intuitive properties of the so called Higgs field.

Most of us are familiar with the "field" terminology from experience with the gravitational field and electromagnetic field. But there are many other fields. Everything in this universe consisting of forces and particles and energies can be represented by fields. Indeed, the whole universe is made up of fields. Peter Higgs, of the University of Edinburgh, first proposed the properties of the field that now bears his name. This field was proposed in an adjustment to the Standard Model of particle physics as a way to account for the mass of particles and the behavior known as "spontaneous symmetry breaking." It turns out that particles such as the electron acquire the property we call mass by an interaction with the Higgs field. There is nothing novel in the idea of proposing an as-yet-unobserved physical entity on the basis of theoretical necessity. The neutrino, the weak-force interaction particles and the recently detected quark, were all proposed on theoretical grounds well before there was any experimental evidence for their existence. The Higgs boson, the force-bearing particle associated with the Higgs field, has not yet been detected. But Higgs fields, and there are several of these, are now so completely interwoven into the fabric of theoretical physics that no one doubts that they exist. Teams at CERN and Fermi Laboratory are racing to be the first to catch a Higgs boson. Respecting this, the headline in the science section of the November 21, 2000, issue of *The New York Times* reads, "Particle Physics Braces for the Next Big Thing." Chances are that the discovery of the Higgs boson will take place by the time these words are published. This particle is so basic to the fabric of existence that Nobel winner Leon Lederman named it the G-d particle (His book is titled *The G-d Particle: If the Universe is the Answer, What is the Question?*). But it takes a great deal of convincing to believe in the Higgs field because it behaves in such a counterintuitive manner.

All other fields exhibit the zero-energy state when the field goes to zero. It seems logical that there is no energy when there is no field. Not so for the Higgs field. This field exhibits zero energy when the field is not zero. I have not come across any analogy or other explanation as to how such a thing is possible. But that is how the mathematics comes out. Perhaps the following suggestion might make the matter more acceptable to the reader. Consider some empty space. We normally call this a vacuum, or the physical vacuum. A true vacuum would be completely empty, devoid of anything. It is a state of lowest energy density. The space of the vacuum is empty, but it is not nothing. The vacuum is not nothing because the space exists. And within this empty vacuum we have a Higgs field and associated Higgs boson. The Higgs field is there at a level which represents the minimum energy density. Remove the Higgs field, and the ordinary vacuum space also disappears. Viewed this way, we could say that the Higgs field maintains the existence of space. The field cannot be zero in the vacuum of space, because the space is empty of everything except itself.

Now comes the interesting part. When we remove something, the Higgs boson, from the vacuum we create something less than the vacuum. This condition is called the "false vacuum." But if the energy density is zero when the Higgs field is some value, then the energy cannot be zero when the field is zero. Indeed, a zero Higgs field corresponds to a large, and I mean huge, energy density. There is enough energy in a zero-level Higgs field to create a whole new universe. Furthermore, while the energy level of the false vacuum is high, the pressure direction is negative. The false vacuum wants to explode into the physical vacuum, releasing all the pent-up energy into a universe-creating bubble. That is what the inflation theory proposes.

At first it was suggested that the zero-field condition resulted in a local energy minimum, which exploded into the true minimum of the physical vacuum by QM tunneling across the intervening energy barrier. But this causes too fast an expansion for a stable universe. This problem was fixed by ascribing a flat energy plateau for the false vacuum, so there is no energy barrier to tunnel through. This permits a "graceful exit" to the inflation, yielding the appropriate conditions for our kind of universe.

Simply put, inflationary expansion starts with a condition of absolute and complete nothing. There is no Higgs field and no physical space, and then ordinary space is created. The reasoning is difficult and counterintuitive. I will not try to confuse the reader any further, and just note that a thorough explanation of this topic is provided in the above-referenced book by Guth. But the matter can also be explained on the basis of conventional QM.

"Conventional QM?" Since when has QM been "conventional?" But still, the QM uncertainty principle is better known and easier to follow than the behavior of Higgs fields. The usual illustrative example of the uncertainty principle is of the combination of position and momentum (velocity). The more accurately we know one, the greater the uncertainty, the fluctuation, of the other. Time and energy also form such a pair. As the time interval gets smaller and smaller, moving down to the Planck time of 10^{-43} seconds, the energy uncertainty gets larger and larger. Presto, you can have lots of energy for a very short time. These fluctuations are unstable and disappear about as quickly as they appear. But all it takes is one that expanded, and we have our universe.

The latent mass-energy of the space-time foam, consisting of quantum fluctuations, is enormous. It has been estimated to be many billions of times greater than the total mass of the observed universe for every cubic centimeter of space. But a cubic centimeter is way out of line with a fluctuation within a space of Planck time and Planck length dimensions at 4×10^{-33} cm. Apparently the initial fluctuation is not enough to provide all that the universe contains. But that is not a problem, because it all adds up to nothing anyway. That is because mass and energy are positive, but gravitation is negative. And guess what? These are precisely balanced. This balance is what, among other things, is meant by a flat universe. Inflation says that we have a flat universe, and the measurements of the cosmic background radiation reported in early year 2000 verify that indeed the universe is flat. The universe is one big something that, when you add it all up, comes to a total of nothing. Nothing is easy to multiply, and that is what happened during the period of inflationary expansion. We can have as much stuff in the universe as we like, or perhaps I should say, as much as Hashem likes. And by the way, the amount just happens to be right to create conditions that are optimal for beings like us.

Not only is the universe optimal for beings like us, but some theories require that we have to exist in order for the universe to exist. This is not a figment of my imagination. This is current science that I discussed in some detail earlier. Here is a quotation of the subheading from an article on this matter in the May, 1987, issue of *Discover* magazine. "The universe exists only because it has been, is being, and will be observed."

I read the above paragraphs, and it seems like nonsense. A nothing which is less than nothing, which makes a something which when you add it all up, is really nothing, and it all came from nothing. And this universe, which is simultaneously something and nothing, exists only because we are around to talk about it. Absent us, and it becomes nothing again, which it actually is anyway. Do I really mean this? Yes, I mean it. But my opinion is not important. What counts is that the highest-level scientists mean it. And to really understand this requires a deeper-level explanation than I have given or am capable of giving. The Torah too seems to be contradicting itself, and to not make sense on a simple reading. We talk about an astonishing and chaotic complexity (*tohu*) and a nothing (*bohu*) which nevertheless contains everything within it (bo = in, hu = it is). A whole book could be written on such items and how science and Torah compare. Indeed, much of Goldfinger's *Thinking About Creation* provides an excellent analysis of such matters. It is a difficult subject, and I have nothing new to contribute here. Fortunately, we need not delve into the details to see what the Torah has to say to us about the beginning of creation. Our Torah Sages have explained it in plain language. Here are a few examples from Maimonides and Nachmanides. Examples from others would follow similar thinking.

Chapter XXX of the second section of Maimonides' *Guide for the Perplexed* explains the following. "The universe has not been created out of an element that preceded it in time – time itself formed part of the creation ...."

"The universe was [created] from nothing … All things were created together" in the beginning, and were later "separated from each other." In the early universe, "the laws of nature were then not yet permanently fixed." Much can be explained about creation, but it is best to stop lest, "I might become a revealer of secrets." In chapter XIII we find the following. "In the beginning G-d alone existed, and nothing else." "[T]ime itself is among the things created." "We say that G-d existed before creation." The verb "existed" implies time before creation. But, says Maimonides, "we do not mean time in the usual sense … but only something analogous or similar to time."

The following statements by Nachmanides were presented at the beginning of this essay. The original creation came from "absolute and total nothing." It was initially like "a very small point having no substance." "Everything" was created at once. The substance in this creation was not the ordinary matter or energy that we now have. Rather, it was "a very thin substance devoid of corporeality but having the power of potency." After the space-time small point and the unique "thin" substance came into being, "He did not create anything, but He formed and made things with it."

Chaotic inflation and Torah

There are a number of early universe inflationary expansion proposals. The choice may change by the time you read this, but chaotic inflation appears to be the current favorite in the year 2001. Possibly this version of the theory has high status because it is proposed by Andrei

Linde, one of the early pioneers on inflation theory. Possibly it is because of how this theory works. Whatever the reasons, this is my choice for further discussion.

Like all inflation variations, it is proposed that the universe came into existence as an expansion of a quantum fluctuation in the false vacuum. And like other inflation scenarios it provides "natural" explanations for various impossible-seeming coincidences inherent in the big bang standard model. But, unlike other inflation versions, chaotic inflation does not call for carefully controlled initial conditions. The initially chaotic quantum foam fluctuations contain all possible starting points, including those favorable to inflation. Furthermore, this scenario starts inflation at the earliest possible time, just about at the Planck time. This is good because the smaller the time interval, the greater the possible energy fluctuation and the less time there is for the inflation to collapse before it has a chance to create a universe. Furthermore, chaotic inflation is conducive to multiple inflation regions side by side. The universe could consist of a multiplicity of such regions, and we would not know it. This is because each inflationary region will appear to be uniform when observed from within the region.

This fits nicely with the weak anthropic principle. Also, differing quantum foam bubbles can yield different, separated universes. This is a nice fit to the strong anthropic principle, as well. And as you will recall, the strong anthropic principle is the best candidate to explain just about everything on a "natural" basis and get rid of any need for a Creator. The strong anthropic principle, combined with chaotic inflation, has got to provide just about the strongest opposition to a need for a Creator and to Torah. Yet I maintain that this does not threaten Torah, and that, in fact, Torah requires this. Why? There are three reasons. Because Torah and science provide remarkably consistent descriptions of the early universe. Because human beings have free will. And because Torah predates modern science by millennia.

The beginning in Torah and science

This is a very large topic. The matter has been considered, to some degree, in previous essays, and several books dealing with this area have been referenced. But much more is available. Here I will just touch upon the elements directly identified in this essay. I am not trying to provide a full or detailed discussion, but to only show that science and Torah are compatible in this regard.

Science is not certain yet as to what happened in the very beginning, and how it all came about. We have a number of competing theories. One can easily disagree with my choices as to what to discuss. But I have endeavored to deal with ideas that seem to hold promise for the future. No doubt, whatever I say will be dated in just a few years. But the general principles should remain. Perhaps a more important factor in my choice, is that I have tried to find science that promises to explain everything on the basis of "natural" means. Torah and a Creator would then seem to be superfluous. If Torah can withstand this, then it should also be able to withstand alternate scientific formulations.

One of the fundamental constants of this universe is known as Planck's constant. It comes about from the discovery by Max Planck that radiation is emitted in bursts, called quanta, which are connected to the frequency by the formula $E = hf$, where $h = 6.63 \times 10^{-34}$ j-sec. This number is known as Planck's constant. Planck's constant, related to energy quanta, shows up in various ways in QM. For instance, the famous QM Heisenberg measurement uncertainty equals $h/2\pi$. A number of minimum-value, so-called constants, related to Planck's constant, show up in QM analysis. Among these are the Planck length equal to 4×10^{-33} cm, and the Planck time of 10^{-43} seconds. The Planck time is the time it takes light to traverse a distance of the Planck length. The normal laws of physics fall apart at such time and distance. Nobody knows what happens as you go smaller. Some theoretical analyses postulate that both time and space are quantized and granular at a sufficiently small time and distance scale. That would mean that the basic fabric of space is discontinuous. In between bits of reality is a nothing which is less than the false vacuum nothing which itself is less than the physical vacuum nothing. Can the Torah support such an idea? Yes, it can. We are told that all reality is being constantly renewed and recreated, and I quoted earlier from Rabbi Dessler about how what we perceive as continuous actually comes to us in granular, quantized jumps. Of course, we cannot see such tiny variations, and the uncertainty principle prevents us from ever being able to make measurements on such a scale. This is because the time and distance involved is not just smaller, but smaller than we can imagine. The diameter of the electron, the smallest standard particle, is 5.6×10^{-13} cm. This means that we can place side by side about 10^{20} Planck lengths inside a single electron. Expanding everything by the same ratio, we find that an electron would be 560 km across, once the Planck length gets to the size of an electron.

The quantum foam in the false vacuum seethes with tremendous energy fluctuations, adding up to many times the total contained in the observable universe in every cubic centimeter of space. But these are confined to Planck-length distances and to Planck-time duration. The time and space is just too small to affect us at our scale of existence. But as the time and space for each foam fluctuation gets smaller, the number of fluctuations gets larger. Irregular, chaotic fluctuations will include some that will expand far enough to experience a phase transition with an enormous release of latent energy. This introduces an accelerated, inflationary expansion, and we suddenly have a new universe. How big was the original universe before it started to quickly expand? It was small, on the order of the Planck length. Or perhaps it was even smaller, involving a dimensionless singularity point. Either way, it fits the Torah description of "a very small point having no substance."

The laws of physics were different before the Planck time. Says Maimonides, "The laws of nature were not yet permanently fixed." Time itself, being granular with a Planck time quantum duration, could not exist in a manner as we now know it. Time, as we now know it, was created by the expansion beyond the Planck length, but something connected to time preexisted this expansion. Explaining time before creation, Maimonides notes that it was "something analogous or similar to time … but not time in the usual sense."

The energy content of the foam fluctuation was not in any way of the type of energy or matter that we now experience. Currently we have hadron-based matter, consisting of gluon-bound quarks in a stable configuration. Prior to that, we had free-energy quanta, and what scientists call a "soup" of free quarks and gluon plasma. This peculiar state of matter has now been verified. See the report titled "Fireballs of Free Quarks," in the April 2000, issue of *Scientific American* magazine. But the transition to bound quarks was way, way after the initial inflationary expansion. Here we are talking about conditions at the very beginning, at the Planck time. The mechanism involves a fluctuation in the false vacuum, an explosion in a nothing which is less than nothing. But this "nothing" eventually became everything. It all started as an "absolute and total nothing," according to Torah, yet within this "nothing" was a "thin" incorporeal substance (not hadron matter) which had the "potency" to yield all we now have. And "He did not create anything" new once the hadron matter we now enjoy solidified. Galaxies, planets, carbon, life, sequence and order and everything we now have came about through "natural"-seeming means after that, as "He formed and made things" with the previously created matter-energy.

We could go on. But the reader should by now see how it goes. The Torah and science use different means and different languages. But the end result is substantially the same. Even the intermediate results and processes are essentially the same. There is only one essential difference here. Science ascribes it all to a "natural" process, while Torah calls for a Creator. In addition, science winds up the natural clock and then lets it continue unhindered. Torah, on the other hand, insists that the universe, and all within it, are under constant control and supervision and all would instantly disappear if it were not so. The scientist is now entitled to ask a simple question: What is the use of Torah, if science can explain everything?

Why Torah, why free will?

A simple answer to the question as to the purpose of Torah if science can explain all by "natural" means, is that Torah is not science. Torah provides us with instructions how to live, a topic that science has nothing to say about. Science tells us what happens, but it does not tell us about the meaning of what happens. Here is how Brian Greene puts it on page 364 of his book *The Elegant Universe*. "That is not to say that modern science provides a connection between the question of how and the question of why – it doesn't – and it may well be that no such scientific connection is ever found." Science does not teach Torah, but Torah does in fact provide us with a most profound teaching of science. It is science for the sake of Torah, and not just as a topic on its own. Only those at the highest level of Torah understanding can be privy to this "science." This is why Maimonides and Nachmanides, as quoted in this essay and numerous other sources, including Tractate *Chagigah*, where the expansion of the universe is discussed, firmly state that certain matters must not be revealed. *Sefer Yetzirah*, which is discussed below, states that matters of creation are "a great mystical secret." These come out only as science catches up with Torah. Thus, a hundred years ago someone of my relatively low status in Torah knowledge could not possibly have fathomed some of the points of Torah I discuss in these essays. I only see the implications of certain Torah items only because I am prepared for it by science. This brings us to an important point respecting the relationship between Torah and science.

There is no science-based methodology to prove this, but we believe that the Torah contains all information, even if much of it is currently hidden from us. As Rabbi Munk puts it on page 47 of his book *The Wisdom in the Hebrew Alphabet*, "The sum total of human knowledge derives from the Torah." In the future, as we learn more science, ordinary people, such as this writer, will also learn more Torah. Let us take a specific example involving the Higgs field, a state which I identify with the "absolute nothing" of Nachmanides. Knowing what to look for, we find a number of modern scientific concepts, including the Higgs field, in the mysterious and most ancient of all Kabbalistic texts, known as *The Book of Creation – Sefer Yetzirah*. Parts of this book are attributed to Abraham, and we are certainly dealing with ideas at least two millennia old. The book content is mysterious and subject to interpretation. But some of the statements are remarkably in line with the latest physics. Permit me a somewhat speculative digression to examine some of these ideas.

The primary subject matter of *The Book of Creation* is the concept of *sefirot*, emanations that are associated with the creation of the universe. It is tantalizing to consider the parallels between the ten sefirot and the ten dimensions of superstring theory. Superstrings, for those who may not be aware, is the current favorite TOE (theory of everything). *The Book of Creation* says that the universe came into existence through sefirot, and science says that the universe came about via superstrings. How do these compare?

We have ten primordial sefirot in connection with the creation of the universe, and we also have ten superstring dimensions. The latest superstring theory, known as M-theory (see *The Elegant Universe*, by Brian Green), tells us that there actually is an additional, eleventh dimension. Similarly, a more sophisticated sefirah structure includes an eleventh so-called "quasi-sefirah." To be sure, there are obvious differences between superstrings and sefirot. The ten sefirot come in pairs of opposites – past/future, good/evil, etc. – and hence define five mathematical lines, or dimensions, and not ten as in superstring theory. But wait a minute. Each of the strings of superstring theory is also a two-dimensional line. Does that make it better? I don't know. Greene explains in his book how all the dimensions in superstring theory started out on an equal basis, and then the four consisting of time and three space dimensions became accessible to us, while the others remain hidden. Kaballah similarly tells us how some sefirot became accessible to us while the others remain hidden. David Peat, in his book *Superstrings and the Search for the Theory of Everything*, tells us on page 321, that bosons, the force-bearing particle, apparently consist of 22 elements, while the mass-bearing fermion is made up of four dimensional space-time plus six other elements for a total of ten superstrings. Is it a coincidence that we have a parallel in the idea that creation consists of ten sefirot plus 22 letters of the Hebrew alphabet? I do not know.

Then we come to the point about the Higgs field I started to make at the outset. In *The Book of Creation*, the highest-level sefirah is identified as the level of *nothingness* (*Ayin* in Hebrew). In his commentary on this book (page 11), Rabbi Aryeh Kaplan explains that, "It is on this level [the level of nothingness] that the laws of nature cease to exist, and can therefore be altered." You can place this statement in any modern physics book, discussing the Higgs field and the nothingness of the false vacuum, and it will fit perfectly. Not only individual elements, but the whole sequence of creation has a parallel description in science and Torah. We thus have the following.

The first level of creation is *Atzilut*, which means an emanation. This is the realm of absolute nothingness. This concept is a well-established element of the hidden knowledge, though it does not appear explicitly in the written Torah. The equivalent in science would be the emanation known as a Higgs field, which is connected to the nothingness of the false vacuum.

Next comes the creation of something from nothing. This is represented by the second word of the written Torah, *barah*. In science we are told that a QM fluctuation in the nothing of the false vacuum brought forth ordinary space-time and energy out of which our universe grew.

The third step in the Biblical sequence is to form something from an existing substance. Symmetry breaking in a phase transition in the inflationary expansion of the early universe formed, or perhaps one should say transformed, the initial QM fluctuation into the gluon-quark mixture that ultimately became ordinary matter.

The last step is known as making and completion. The establishment (the completion) of the current laws of the universe came about as the temperature dropped to the point where the hot gluon-quark plasma "froze" into the final form of hadron-based matter that the universe currently consists of.

Parts of this book have been with us for over two thousand years. Would someone of my limited knowledge of Torah have understood its implications 500 years ago? Certainly not. It is only owing to my knowledge of modern science that I can discern certain meaning in this book. But this information, and other things that I now do not understand, has been in the book from its inception. It is a given for those who accept the Torah that science will never outstrip Torah. We will find Torah-based counterparts for whatever scientific truths and theories are discovered in the future. And the Torah will also address the "why" as well as the "how."

Once upon a time, a primitive science appeared to threaten Torah. But no more. Today, contrary to popular misconceptions based on the past, it is not Torah that is threatened by science. Rather, it is science that is threatened by Torah. Many scientists have recognized this fact, stating that certain scientific facts clearly point to a Creator. Others have reacted with an all-out attack on Torah as the enemy. Whatever the reaction, as I have said, it is science and not Torah that is threatened, now that science has abandoned early simplistic notions about the universe. But science is not standing still. Today, as science has made progress, it would appear that a "natural" explanation might one day be found for all observed phenomena, and science will have gotten rid of Torah, once and for all. Or will it?

Torah absolutely insists that human beings have free will. This creates a difficult paradox, as the idea of an omniscient and omnipotent Creator – who is actively involved in the day-to-day affairs of the universe in general, and each and every human being in particular – is juxtaposed with the idea that we are free to make choices and act upon them. Numerous books on Torah deal with this matter, and we discussed this in previous essays. All I will say here is to reiterate that we have free will. Thus, Rabbi Luzzatto states, in *The Way of G-d*, that we have "the power of choice." Humanity differs "from all other species, since it was given free will." Commenting on the Mishnah *Pirke Avot*, Maimonides emphasizes that "we are convinced" on the basis of proofs from the Torah that "man's conduct is entirely in his own hands." Among his proofs is the oft-repeated statement that, "All is in the hands of Heaven except the fear of Heaven." The reader will find the details in Tractates *Berachos* 33b and *Megillah* 25a. Respecting the requirements of Toràh, "man has absolutely free choice to perform them or not to perform them." And we also have a free choice in what to believe.

We can choose to believe that the Torah is true or that it is false, and we can ignore the existence of the Torah altogether. This provides one reason that it should be possible for us to explain the workings of the universe as part of a "natural" process devoid of a Creator. Anything else would compel us to accept the truth of Torah. We are different from all other created beings in this regard (see quote from Luzzatto above), the essential difference being that we are created in the (mental capacity) image of the Creator. The statement I quoted from Hawking, that science compels him to consider a Creator, is a sign of an immature science. Torah indicates that ultimately science will no longer compel such a conclusion. It would appear that the Torah requires that science will one day permit a completely "natural" explanation for the workings of the universe. This applies even in the direct presence of a miracle, because as we are told in *Niddah*, page 31a, that "one who experiences a miracle does not recognize it." But Torah also assures us that it too contains all of the science-derived truths that we will one day discover. Science will then no longer be threatened by Torah, and Torah will not be asked to defend itself against science.

We are free to choose as we wish. We can choose a false vacuum fluctuation or *barah* of a Creator. We can accept a Creator who maintains the universe on a continual basis, or we can believe in a strong anthropic principle that suggests that we ourselves make the universe real through the act of measurement and observation. We can be satisfied with an inflation theory proposed by Guth, or we can delve into the explanation given to us by Resh Lakish. These would appear to be equally valid, except for one matter. Guth is a contemporary of ours, while Resh Lakish lived almost two millennia ago. How did he know about an expanding universe? "See, I have set before you this day …, therefore choose" (Deuteronomy 30:15). We have free will, and we can freely choose, just as the Torah tells us.

References

This book contains quotations from, or reference to the following books:

Aiken, Lisa. ***To Be A Jewish Woman***. Jason Aronson, 1992.

Aiken, Lisa. ***Why me G-d? A Jewish Guide for Coping with Suffering***. Jason Aronson, 1997.

Barrow, John and Frank Tipler. ***The Anthropic Cosmological Principle***. Oxford University Press, 1986.

Carmel, Aryeh and Cyril Domb. ***Challenge***. Feldheim Publishers, 1976.

Chavel, Charles. ***Ramban (Nachmanides) Commentary on the Torah***. Shilo Publishing, 1971.

Darwin, Charles. ***The Origin of Species***. Grammercy, 1998.

Dawkins, Richard. ***The Blind Watchmaker***. Penguin and Norton, 1987.

Dessler, Eliyahu. ***Strive for Truth***. Rendered into English by Aryeh Carmel from *Michtav Me-Eliyahu*. Feldheim Publishers, 1985.

Feynman, Richard. ***The Character of Physical Law***. The M.I.T. Press, 1965.

Feynman, Richard as told to Ralph Leighton. ***Surely You're Joking Mr. Feynman***. Bantam Books, 1985. Numerous more recent editions, such as Norton, 1997.

Goldfinger, Andrew. ***Thinking About Creation***. Jason Aronson, 1999.

Greene, Brian. ***The Elegant Universe***. Vintage Books, 2000.

Gribbin, John and Martin Rees. ***Cosmic Coincidences***. Bantam Books, 1989.

Grunfeld, Isidor. ***The Sabbath***. Feldheim Publishers, 1954.

Guth, Alan. ***The Inflationary Universe***. Perseus Books, 1997.

Hawking, Stephen. ***A Brief History Of Time***. Bantam Books, 1988.

Kaplan, Aryeh. ***Immortality, Resurrection, And The Age Of The Universe: A Kabbalistic View***. KTAV Publishing House, 1993.

Kaplan, Aryeh. ***Translation and commentary on The Bahir***. Samuel Weiser, 1979.

Kaplan, Aryeh. ***Translation and commentary on The Book of Creation (Sefer Yetzirah)***. Samuel Weiser, 1997.

Kelemen, Lawrence. ***Permission To Believe***. Targum Press & Feldheim Publishing, 1990.

Kornfeld, Mordecai. ***Torah From The Internet***. The Judaica Press, 1998.

Lederman, Leon and Dick Teresi. ***The G-d Particle: If the Universe is the Answer, What is the Question***. Delta, 1994.

Luzzatto, Moshe Chayim. ***The Way of G-d***. Feldheim Publishers, 1977.

Maimonides, Moses. ***The Guide for the Perplexed***. Friedlander translation. Dover Publications, 1956.

Munk, Michael. ***The Wisdom in the Hebrew Alphabet***. Mesorah Publications, 1983.

Pagels, Heinz. ***Perfect Symmetry***. Bantam Books, 1985.

Paley, William. ***Natural Theology: Evidence of the Existence and Attributes of the Deity: Collected from the Appearance of Nature***. Classworks, 1986.

Peat, David. ***Superstrings and the Search for The Theory of Everything***. Contemporary Books, 1988.

Rees, Martin. ***Just Six Numbers***. Basic Books, 1999.

Ribiat, Dovid. ***The 39 Melochos***. Feldheim Publishers,1999.

Ryan, William and Walter Pitman. ***Noah's Flood: The New Scientific Discoveries About The Event That Changed History***. Simon & Schuster, 1999.

Schroeder, Gerald. ***The Science of G-d***. Broadway Books, 1997.

Talmud and Torah. Several translations have been used in support of the original, which is the primary reference. This author prefers the Artscroll translation, published by Mesorah Publications.

Tatz, Akiva. ***Worldmask***. Targum Press, 1995.

Twerski, Abraham. ***Let Us Make Man***. CIS Publication, 1987.

Ward, Peter and Donald Brownlee. ***Rare Earth***. Copernicus Books, 2000.

Weinberg, Steven. ***The First Three Minutes***. Basic Books, 1988.

The following internet sites are noted in this book.

On Torah. www.dafyomi.co.il ; www.shemayisroel.co.il

On the temperature structure of the cosmic microwave background radiation. www.physics.ucsb.edu/~boomerang/press_images/index.html

On Olbers paradox. www.zabu.uoregon.edu/~imamura/209/apr7/olbers.html

Quantum teleportation. http://info.uibk.ac.at/c/c7/c704/qo/photon/_teleport/index.html.

On quantum mechanics and other advanced science matters. www.npl.washington.edu/av

The cover image comes from the Sloan Digital Sky Survey project. The reader will find this, and other images at: www.sdss.org

Index

About the author

Morris Engelson was educated as an electronics engineer specializing in spectrum analysis and communications theory, and elected a Fellow of the Institute of Electrical and Electronics engineers "for contributions to the practice and applications of spectrum analysis." He is a former Chief Engineer at Tektronix Inc., and Adjunct Professor at Oregon State University. The author of several books and a number of technical papers in his specialty, Morris Engelson is currently a consultant located in Portland, Oregon. You can reach the author by email at:

m.engelson@ieee.org